AF870242

Resilience of Computer Systems

Igor Schagaev

Resilience of Computer Systems

Resilient and Evolving Computer Systems Design

Second Edition

Springer

Igor Schagaev
Applied Computer Science and Technology
IT-ACS Ltd.
London, UK

ISBN 978-3-032-00937-1 ISBN 978-3-032-00938-8 (eBook)
https://doi.org/10.1007/978-3-032-00938-8

1st edition: © Springer International Publishing Switzerland 2015
2nd edition: © The Editor(s) (if applicable) and The Author(s), under exclusive license to Springer Nature Switzerland AG 2026

This work is subject to copyright. All rights are solely and exclusively licensed by the Publisher, whether the whole or part of the material is concerned, specifically the rights of translation, reprinting, reuse of illustrations, recitation, broadcasting, reproduction on microfilms or in any other physical way, and transmission or information storage and retrieval, electronic adaptation, computer software, or by similar or dissimilar methodology now known or hereafter developed.
The use of general descriptive names, registered names, trademarks, service marks, etc. in this publication does not imply, even in the absence of a specific statement, that such names are exempt from the relevant protective laws and regulations and therefore free for general use.
The publisher, the authors and the editors are safe to assume that the advice and information in this book are believed to be true and accurate at the date of publication. Neither the publisher nor the authors or the editors give a warranty, expressed or implied, with respect to the material contained herein or for any errors or omissions that may have been made. The publisher remains neutral with regard to jurisdictional claims in published maps and institutional affiliations.

This Springer imprint is published by the registered company Springer Nature Switzerland AG
The registered company address is: Gewerbestrasse 11, 6330 Cham, Switzerland

If disposing of this product, please recycle the paper.

Preface I

There are several reasons of making second edition of our book *Resilient Computer System Design*, ISBN 13: 9783319150680.

At first, the objective—since the first edition physics was not denied or changed and all our research about the impact of radiation on electronics is valid and even became more important. It is worth mentioning that our research especially in this segment was appraised by NASA Prognostic Centre Director Dr. K. Goebel.

Surprisingly, since then nothing tangible has been made regarding design of computers for safety-critical applications and, regretfully, neither analysis as it should nor developments that might change the situation are known so far. Thus, we (I mean myself and colleagues: Dr. Victor Castano, Prof. Juerg Gutknecht) became obliged to present new development and trim already published edition.

Add to this that our own processor that best fit for safety critical systems in terms of requirements and each instruction of which was tested and verified by Prof. Hao Kai, (now Dean of Computer Science Department in Shantou University), we believe that our concepts and solutions in both—theoretical and technological aspects were proven and applied.

The whole material of the first edition was heavily edited, all references and links updated and checked.

Second edition became enriched by several new chapters. Thus, our long discussions and friendly chats with Prof. Juerg Gutknecht (ETHZ) about what kind of interaction between entities is required and how to represent them at the level of a problem, program and hardware—in terms of interest of this book—organization of interruption were finally summarized and form Chap. 8. Our concept of fault-tolerant semaphores (FTS) was not only justified by our experience and system programming but also illustrate as hardware solution that any good M.Sc. student might repeat. It means that future systems will be, indeed, designed using FTS.

We construct a conclusion of our book as a chapter deliberately, as this chapter presents not only what we have analyzed and done but highlighting what is required to develop, investigate and apply further.

Further in Chap. 11, we extended our analysis and comparison with well-promoted Berkley view on future computer architectures having our own arguments, examples and solutions. It is much more should be done and developed in both—hardware and system software to make our safety-critical resilient system efficient and practical. This chapter is based on our own practice, projects and thoughts, and presents structures of existing systems in which we were involved as designers and engineers.

Finally, Chap. 12 of this book is a kind of reflection on non-functional requirements at large from the phases of system concept and design. Resilience, as well as fault tolerance, reliability or any other Non-Functional Requirement (NFR) should be considered as this chapter suggests. Very good point here that we present real example of the devices that were made in concert with design principles of pursuing the degree of freedom—this way we are summarizing our book from the point of view of design engineering.

London, UK
February 2025

Prof. Igor Schagaev

Preface II

This book is the second edition of our work at IT-ACS LTD on resilient systems. We have tried to be consistent and concentrate primarily on development of our paradigm of evolving computers. This also includes methods of analysis and synthesis of evolving computer systems.

We also discuss the abilities of system to match changing requirements and ability to tolerate internal hardware faults, with supportive analysis of technological advances and drawbacks.

We did this book as an essential working material for ourselves in terms of "all you need to know to design and analyze new generation of computer systems" addressing in non-mutually exclusive way reliability, fault tolerance, performance and resilience properties of hardware.

We have introduced and developed supportive models for main hardware elements: processors, memories and interfaces.

We were thinking about the new architecture to fit the following market:

1. Safety-critical
2. autonomous
3. real time
4. military, banking terminal
5. wearable
6. health care systems.

Proposed approach was developed up to hardware prototype—special chapters that describe it demonstrate at the order of magnitude higher efficiency in comparison with existing systems.

Who are our segments of society that might benefit from reading our book and why? Research community will get a consistent description of the theoretical area for further developments.

Industries of hardware and system software designs, manufacturing and exploitation will get pathways to make performance, reliability and energy-smart systems with consistency, enabling unification of market of consumer electronics, safety-critical, embedded, autonomous and autonomous systems.

ICT consumers will get much higher efficiency (and value for money) from their systems (if, of course, devices and systems will be designed according to the principles proposed in the book).

Also, our book can be helpful for the reader in terms of:

1. Analysis of existing systems is given in essence, showing how "classic" solutions stand and work.
2. Existing technological drawbacks are clarified and presented consistently, with proposed solutions that "best fit the requirement" of new computer system.
3. Description of a process of introduction of new properties as a framework for the next generation of
4. Computer system enabling a reader to make consistent analysis of—we stress—all possible system design solutions.
5. Described prototype of evolving reconfigurable architecture might be attractive for students and hardware design engineers, as they, through the book, will discover that computers might be designed much simpler, power efficient and at an order of magnitude more reliable, especially where continuous operation is of utmost importance and failure can be profound.
6. Analysts who have discovered that the market dominance of the general computing systems has been now limited by appearance in billions of embedded systems might find this book useful in terms of merging these market segments.
7. Any reader will be able to use trail simulator and start programming for new architecture.

Nowadays, radiation becomes a serious threat to the reliable operation of safety-critical systems. Fault avoidance techniques, such as radiation hardening, have been commonly used in space applications.

However, hardened components are expensive, lag behind commercial components in performance and do not provide 100% fault elimination. Without supportive structural solutions to provide fault tolerance, hardware faults become system errors at the application or system level, which in turn can result in catastrophic failures.

Through the whole our books and projects, we read the fault tolerance and resilience as processes, instead of static properties. We analyze the physics of radiation-induced faults, the damage mechanisms of particles and the error process.

We suggest new approach to hardware and system software design, combining efficiently for reliability, performance and power consumption for the system.

We demonstrated how new properties of the computer system will be implemented; we introduced a new conceptual system element called a syndrome.

We describe how syndrome can be applied to monitor performance, reliability and energy-smart operations of hardware. Implemented by hardware and supported by system software, syndrome serves as a core of a resilience of architecture.

Using syndrome enables the system (through software and hardware) to be adaptable to:

1. various and modifiable functional requirements
2. different internal conditions
3. environmental impacts

We implemented a software simulator and disassembler and introduced a testing framework to our evolving reconfigurable architecture, inviting researchers and engineers to put "hands on" new computer system with unique properties.

However, these components are expensive, lag behind commercial components with regard to performance and do not provide 100% fault elimination. Without fault tolerant mechanisms, many of these faults can become errors at the application or system level, which in turn, it can result in catastrophic failures. To address these problems and create a kind of ground to make resilient systems we:

1. Studied the concepts of fault tolerance and dependability and extend these providing our own definition of resilience. We analyzed the physics of radiation-induced faults, the damage mechanisms of particles and the error process.
2. Provided extensive taxonomies of (1) existing fault-tolerant techniques and of (2) the effects of radiation in state-of-the-art electronics, gathering and comparing their characteristics.
3. Proposed a detailed model of faults and provided a classification of the different types of faults at various levels. We introduced an algorithm of fault tolerance and define system states and actions necessary to implement it.
4. Introduced novel hardware of both—processor and memories deliberately designed for resilient computer system, and analyzed their efficiency in terms of performance and reliability.
5. Developed system software techniques that provide an efficient combination of reliability, performance and power consumption than known.
6. Proposed a new element of the system called syndrome that is the core of a resilient architecture whose software and hardware can adapt to reliable and unreliable environments.
7. From the system level concept of interaction down to implementation, we clarified what kind of interruption system must be designed for a resilient computer system, showing essential schemes of new logic structures—so-called T-logic.
8. Implemented a software simulator and disassembler and introduced a testing framework in combination with ERA's assembler and commercial hardware simulators.
9. Presented what is still required to do in research and developments in the field of resilience to make better systems.
10. We have also presented our vision for the future computer architectures, comparing with Berkley approach.
11. Philosophical principle of design based on pursuing a high degree of freedom is also presented, using our own designs as examples.

Book Structure

The book is divided into 12 chapters:

Chapter 1. *Introduction:* The chapter summarizes the problems of safety-critical domain of computing, highlighting our approach.

Chapter 2. *Resilience:* We provide part of the theoretical framework of reliability. We analyze the properties of classic dependability and we describe our own view of the concept of resilience.

Chapter 3. *Dealing with Faults: Redundancy:* This chapter provides a complete review of state-of-the-art techniques employed to deal with faults and explores the different types of redundancy and fault-tolerant techniques.

Chapter 4. *FT Models of Fault Tolerance:* This chapter describes the physical mechanisms of radiation as the primary phenomenon that causes faults in current computing systems. We also analyze their effect on semiconductors at low, circuit and system levels.

Chapter 5. *Classification of Faults by Manifestation, GAFT:* We analyze a model of hardware faults. We introduce GAFT and define the different states and actions required to implement fault tolerance.

Chapter 6. *Hardware Support of Resilience:* This chapter details the hardware and system software elements of a novel resilient architecture that can achieve required levels of performance, reliability and energy consumption.

Chapter 7. *System Software Support for Resilience:* This chapter focuses on the development and testing of the hardware prototype. Details of the design and development of a software simulator of the hardware architecture are also provided.

Chapter 8 presents new development of interruptions, schemes that fit way better existing to achieve resilience of hardware and a system as a whole.

Chapter 9 presents our approach, methods, models, instruments and own hardware prototype with required specifications, explaining principles and ways to build properly next generation of resilient systems. Here we also compare our vision about future of computer systems and compare it with Berkley approach. Shown weakness and drawbacks of the latter.

Chapter 10 is a kind of summary our "practical attempts" to implement own and known concepts of resilience.

Chapter 11 presents our vision on future of evolving systems resilient systems discussion new approaches and models we propose, illustrating by the way drawbacks of well-promoted "vision of next generation of system design."

Finally, Chap. 12 presents a kind of philosophical principle in design of human-made systems, presenting our own solutions as proof that it is possible.

London, UK Prof. Igor Schagaev

Contents

1 Introduction 1
1.1 Motivation 1
1.2 Scope and Contribution 4
References 6

2 Resilience 9
2.1 System Failure Lifecycle 9
2.2 Attributes and Measures of Resilience 11
2.3 Reliability 12
2.3.1 Performance and Reliability 12
2.3.2 Reliability and Unreliability Functions 15
2.3.3 Probability Density Function 16
2.3.4 Failure Rate Function 18
2.3.5 Cumulative Hazard Function 19
2.3.6 Bathtub Curve of Failure Rates 19
2.3.7 Mean Time Between Failures (MTBF) 21
2.3.8 Mean Time to Failure (MTTF) 22
2.3.9 Reliability Prediction 23
2.4 Safety 28
2.5 Security 29
2.5.1 Integrity 29
2.5.2 Maintainability 29
2.5.3 Availability 33
2.6 Performability 37
2.7 Resilience 38
2.7.1 Requirements 39
2.7.2 Effectiveness of Resilience 40
2.8 Conclusion 41
References 42

3 Dealing with Faults: Redundancy 45
3.1 Handling Faults: Design Strategies 45
3.2 Fault Avoidance 46
3.3 Fault Tolerance: Using Redundancy 48
3.3.1 Redundancy Notation 49
3.4 Structural Redundancy HW(*S*) 51
3.4.1 Static Redundancy 52
3.4.2 Dynamic Redundancy 57
3.4.3 Hybrid Redundancy 61
3.5 Information Redundancy 64
3.5.1 Error Detection Codes: EDC 65
3.5.2 Error Correction Codes: ECC 66
3.6 Time Redundancy 72
3.6.1 Concurrent Error Detection: Basics of Time Redundancy 73
3.6.2 Alternating Logic 76
3.6.3 Recomputing with Shifted Operands (RESO) 77
3.6.4 Recomputing with Rotated Operands (RERO) 78
3.6.5 Recomputing with Swapped Operands (RESWO) 79
3.6.6 Recomputing with Comparison (REDWC) 80
3.7 Comparison of Main Redundancy Schemes 80
3.8 Conclusion 81
References 82

4 FT Models of Fault Tolerance 87
4.1 Models 87
4.2 Model of Fault 92
4.3 Classification of Faults by Origin 93
4.3.1 Level Response 93
4.3.2 Cause of Faults 94
4.3.3 Phase of Formation and Occurrence of Faults 96
4.3.4 Nature/Dimension 96
4.3.5 System Boundaries 96
4.3.6 Phenomenological Cause 96
4.3.7 Capability/Objective/Intent 97
4.4 Conclusion 97
References 98

5 Classification of Faults by Manifestation, GAFT 101
5.1 Introduction 101
5.2 Response-Timeliness 101
5.2.1 Consistency 104
5.2.2 Maintainability: Detectability, Diagnosability and Recoverability 105
5.3 FT and System Modelling 111
5.4 Trading P, R, E 111

5.5 GAFT: Generalized Algorithm of Fault Tolerance 113
5.5.1 GAFT: System States and Actions to Implement Fault Tolerance 117
5.6 Conclusion 119
References 120

6 Hardware Support of Resilience 123
6.1 Introduction 123
6.2 ERA Concept, System Design and Hardware Elements 125
6.3 ERA Hardware Configuration: ERA 127
6.3.1 Active Zone 127
6.3.2 ERA Processor Architecture 131
6.3.3 Instruction Set 135
6.4 Architecture Comparison and Evaluation 136
6.4.1 Processor Architectures: The Overview 138
6.4.2 Passive Zone 143
6.4.3 Interfacing Zone 143
6.5 ERA Reconfigurability 146
6.5.1 T-Logic for Memory Management 146
6.5.2 T-Logic for Configuration in ERA 148
6.6 Syndrome 149
6.6.1 Syndrome Use 149
6.6.2 Location Access and Way of Operation of the Syndrome 154
6.6.3 Syndrome: Passive Zone Configurations 156
6.7 Graceful Degradation 159
6.7.1 Graceful Degradation: Markov Analysis 161
6.8 Implementation Constraints 162
6.8.1 Graceful Degradation: Solutions for Memory 163
6.8.2 Interfacing Zone: The Syndrome as Memory Controller 164
6.8.3 Access to the Syndrome 165
6.9 Conclusions 166
References 167

7 System Software Support for Resilience 169
7.1 Introduction 169
7.2 System Software Support of Hardware Checking 170
7.3 FT Scheduling 173
7.4 System Software Support for Hardware Reconfiguration 174
7.4.1 The General Procedure of System Software Support 176
7.5 System Software Monitor of Hardware Condition 177
7.6 Conclusion 179
Reference 180

8 On Interruption ... 181
8.1 Interruption as an Interaction ... 181
8.2 Organisation of Mutual Exclusion ... 183
8.3 Concurrency, Further Steps: Fault-Tolerant Interactors ... 186
8.4 Conclusion ... 192
Appendix 1: T Logic True Table ... 192
Appendix 2: T Logic True Table Source Code ... 194
References ... 194

9 Implementation: Hardware Prototype, Simulation and Testing ... 197
9.1 Introduction ... 197
9.2 Instruction Execution ... 198
9.3 Instruction Set ... 201
9.4 ERA Hardware Prototype ... 203
9.5 Conclusion for Chapter ... 206

10 Conclusions for the Book so Far ... 207
10.1 What We Have Done so Far ... 207
10.2 Next Steps ... 212
References ... 213

11 Vision on Evolving System Future ... 215
11.1 Fundamental Problem ... 215
11.2 Known Solutions (What We Have...) ... 216
11.3 Attempts to Evolve ... 218
11.4 Proposed Approach (What We Need and Why We Need This) ... 223
11.5 Supportive Models ... 225
11.5.1 Control-Data-Predicate (CDP) Model ... 225
11.5.2 Graph Logic Model (GLM) ... 227
11.6 Coverage—Generalized Model ... 230
11.7 System Software for Evolving Systems ... 233
11.7.1 Active Language (AL) ... 234
11.7.2 Active Reconfigurable Run-Time System ... 236
11.8 Evolving System: Hardware ... 239
11.8.1 Basic Schemes ... 239
11.9 Evolving System: Multi-element Configuration ... 242
11.10 Evolving System Approach Versus Berkley View ... 244
11.11 Evolving System: Conclusion ... 244
References ... 247

12 Some Further Notes: Degree of Freedom ... 249
12.1 Bits on Degree of Freedom ... 249
References ... 255

Index ... 257

About the Author

Prof. Igor Schagaev 1978 was involved in the projects of Navy computer system, fault-tolerant parallel satellites computer systems, safety-critical real-time computer systems for both—commercial and military aircrafts. All project were about news systems—primarily safety-critical ones, with new hardware design and special system software. Through his career, Igor did consult and collaborated with Sukhoy Design Bureau, British Aerospace, Boeing and Airbus. New hardware and system software, as well as a new theory of active system control were results of this life-long journey: British Aerospace (Filton) in 1994–96, Boeing (Seattle HQ, both—military and commercial aircraft departments, 1998–1999) were involved in joint intensive collaboration. Later, leading NASA Centre of diagnostic scientists were consulted by Igor and joint work was presented twice at leading joint FlighGlobal Forums, International air fleet exhibitions. British Aerospace specialists were following the keynote speeches and my sessions of hardware and system software design at World-Comp in USA for 7 (seven) years 2011–2017. Practically, New hardware and system software design for resilient systems were a main part of the ONBASS project funded by EC DG Research. Key models to design these systems were patented in 2006. Advanced Concept Development of Philips Medical Systems were requesting Prof. Schagaev team consultancy on hardware design for their systems. Same IE Electronics—supplier of Nokia. During the whole period of work in this field, the most distinguished scientists and engineers were involved and supporting Igor works: Turing Award Laureate Prof Niklaus Wirth, Prof. Alessandro Birolini EC DG ICT Prof. Metakides Prof. Bernard Courtois. Great support from two EC VP: Dr. Janez Potocnik and VP EC Neil Kinnock also worth to mention.

Aerospace Editors of *Wall Street Journal*, *Times* were visiting London and Brussels over these years making interviews regarding works of Professor team, seeking some advice on existing aviation hardware and system reliability problems. Facultimedia (Boston), British Computer Society, Institute of Analysts and Programmers were requesting keynote speeches on their congresses and conferences—primarily on the subject of this book. Thus resilient hardware design described in our book is based on exceptional theoretical results and practical experience at the edge of application domain.

Nomenclature

ARQ	Automatic repeat quest
ASIC	Application specific integrated circuit
ASW	Application software
ATPG	Test pattern generation tools
BCH	Bose Chaudhuri Hocquenghem
BEC	Backward error correction
BICMOS	Bipolar complementary metal oxide semiconductor
BIST	Built In Self Test
BPSG	Boron phosphor silicate glass
CCD	Changed coupled device
CED	Concurrent error detection
CM	Corrective maintenance
CMF	Common mode failure
CMOS	Complementary metal oxide semiconductor
COTS	Commercial off the shelf
CSP	Cold standby spare
CUT	Circuit under test
DDD	Displacement damage dose
DEC/TED	Double bit error correction and triple bit error detecting
DFT	Design for testability
DMR	Dual modular redundancy
DRAM	Dynamic random access memory
DRE	Detected recoverable error
DUE	Detected unrecoverable error
DUT	Device under test
DW	Data word
ECC	Error correcting codes
EDAC	Error detection and correction codes EDC Error detecting codes
EEPROM	Electrically erasable programmable read only memory
EPI	Epitaxial substrate doping
FCR	Fault containment region

FEC	Forward error correction
FIT	Failures in time
FM	Fault model
FPGA	Field programmable gate array
FT	Fault tolerant
GAFT	Generalized algorithm of fault tolerance
GCR	Galactic cosmic ray
GDS	Gracefully degrading system
HARQ	Hybrid automatic repeat request
HSP	Hot standby spare
HW	Hardware
ICV IDDQ	Checkable voter
IDDQ	Quiescent power supply currents
IDE	Integrated Development Environment
Iff	If and only if
IR	Information redundancy
LET	Linear energy transfer
MBU	Multiple bit upset
MCU	Multiple cell upset
MOS	Metal oxide semiconductor
MOSFET	Metal oxide silicon field effect transistor
MSB	Most significant bit NIEL Non ionizing energy loss
MTBF	Mean time between failures
MTTD	Mean time to detection
MTTF	Mean time to failure
MTTR	Mean time to repair/restore
nMOS	N channel metal oxide semiconductor
NMR	N modular redundancy
ORA	Output response analyzer
PCSE	Power cycle soft errors
PDF	Probability density function
PI	Primary input
PKA	Primary knock on atom
PM	Preventive maintenance
pMOS	P channel metal oxide semiconductor
PSF	Pattern sensitive fault
REDWC	Recomputing with comparison
RERO	Recomputing with rotated operands
RESO	Recomputing with shifted operands
RESWO	REDWC Recomputing with swapped operands
RF	Register file
ROM	Read only memory
RS	Reed Solomon
RT	Real time RTS Real time systems
RTS	Real time systems

SAF	Stuck at fault
SBU	Single bit upset
SDC	Silent data corruption
SEBO	Single event burnout
SEC DED	Single error correction and double error detection
SEDR	Single event dielectric rupture
SEE	Single event effect
SEFI	Single event functional interrupt
SEFLU	Single event fuse latch upset
SEGR	Single event gate rupture
SEHE	Single event hard error
SEL	Single event latch up
SEMU	Single event multiple upset
SER	Single event rate
SESB	Single event snapback
SET	Single event transient
SEU	Single event upset
SOC	System on a Chip
SOI	Silicon on insulator
SOS	Silicon on sapphire
SPF	Single point of failure
SR	Structural redundancy
SRAM	Static random access memory
SSW	System Software
SW	Software
TID	Total ionizing dose

Chapter 1
Introduction

Abstract Embedded systems occupied the human world in millions. Extreme, safety-critical applications demand tolerance to radiation, highest reliability, flexibility in performance, and reconfiguration. All these require handling parallelisation, efficiency of energy consumption.

Keywords ICT · Radiation · Embedded systems · Performance · Reliability · Energy trade

1.1 Motivation

Embedded systems are ubiquitous nowadays, built into homes, offices, bridges, medical instruments, cars, airplanes, and satellites and even into clothes. The market size of such systems is already larger than the one for general purpose computing. The majority of embedded systems are real-time systems (RTSs) and mostly all RTSs are embedded in a product.

For decades, embedded RTSs are being used in fields where their correct operation is vital to ensure the safety and security of the public and the environment: from automotive systems and avionics to intensive health care and industrial control as well as military operations and defense systems. These systems are subject to time constraints and must guarantee a response within specified timing bounds. The safety critical nature of RT embedded systems employed in those fields demands the highest possible availability and reliability of system operation.

The exponential growth of clock frequency and memory size has led to important achievements in the technological development of microprocessors. Manufacturers of advanced silicon electronics have been able to create more complex designs by periodically scaling down the technology, increasing the transistor density. This growth is supported by the progressive miniaturization of electronic components predicted by Moore's law in 1965.

This phenomenon has also produced undesirable consequences that introduce physical limitations to the law. Due to the area reduction of electronic components to

© The Author(s), under exclusive license to Springer Nature Switzerland AG 2026
I. Schagaev, *Resilience of Computer Systems*,
https://doi.org/10.1007/978-3-032-00938-8_1

nanometer scales and due to the increase in clock frequencies (ITRS 2011), supply voltages have been reduced to keep power dissipation manageable while thermal noise voltages have increased (Asanovic et al. 2006; Kish 2002).

For a long time, radiation effects have been a serious concern in aviation and spacecraft electronics. As the dimensions and voltages of embedded systems are reduced, their sensitivity to ionizing particles has considerably increased. Energizing particles can produce a number of faults at the hardware level, not only in contexts with harsh environmental conditions, such as outer space, but also at sea level with regular conditions.

Components with lower power and noise margins are less reliable and therefore recent systems are more prone to transient faults induced primarily by radiation (Baumann 2002, 2005; Seifert et al. 2002; Shivakumar et al. 2002).

Transient faults do not cause permanent damage in circuits but can affect system behavior by corrupting stored information or signal communication (Mavis 2002; Mavis and Eaton 2002).

Besides, the typical stress experiments in laboratories based on particle bombarding, there is a considerable amount of evidence of radiation induced malfunctions and catastrophic failures during operation in real-life environments.

Radiation-induced faults are frequent in space environments (Adams and Gelman 1984; Adams et al. 1982; Binder et al. 1975; Blake and Mandel 1986; Blandford et al. 1984; Waskiewicz et al. 1986). The Saturn's Cassini (Swift and Guertin 2000), Mars Odyssey and Jupiter's Galileo (Fieseler et al. 2002) are examples of missions that presented malfunctionsas a result of cosmic rays.

The satellites X-ray Timing Explorer (Poivey et al. 2004), Gravity Probe B (Owens et al. 1996), TOPEX/Poseidon (Johnston and Swift 1997) and GRACE (Pritchard et al. 2002) have also reported anomalies during operation.

Radiation Induced faults are also present to a lesser extent in atmospheric (Taber and Normand 1992, 1993) and terrestrial environments (Hauge et al. 1996; Normand et al. 2010; Ziegler 1996).

As a result of this scenario, there is an increasing need to deal with faults. There are two classes of mechanisms to deal with them: fault avoidance and fault tolerance FT (Laprie et al. 2004). Fault avoidance means developing components/systems that are less likely to present faults while fault tolerance techniques focus on the system's ability to tolerate the effects of these faults.

Fault tolerance is defined as the *ability to provide uninterrupted service, conforming to the desired levels of reliability even in the presence of faults* (Laprie et al. 2004). Applications of modern electronic systems require more and more mechanisms to mitigate the effect of these faults (Nicolaidis 2010).

Complete avoidance of faults in a system is practically impossible, and hence, a balance of the two approaches is currently applied. With the objective of reducing costs, attempts to apply fault tolerance (FT) to COTS computers did not bring any substantial breakthrough neither in efficiency nor any other crucial property required from safety-critical systems (Antola et al. 1986).

The research community mainly focuses on

(a) identifying all possible mechanisms leading to accidents and
(b) on providing pre-planned defense techniques against them.

However, too little attention to potential systems that can respond to deviations from desirable states has been paid.

The research is driven by observations on resultant limitations from the evolution of computer architectures, which have been motivated by technological and market choices as well as physical limitations.

These observations refer to performance deceleration, increase in power consumption/dissipation, reliability aspects, parallelization challenges, design complexities, and hardware and software inefficiencies. A brief explanation for each follow:

Performance deceleration: As transistor density and frequency increase to satisfy the immediate market demands, unjustified complexity has been introduced in the current computer architectures. In recent years, clock rates of commodity microprocessors have flattened and performance of processor cores has slowed down (Asanovic et al. 2006; Hill 2001).

Power consumption/dissipation: Scaling processor clock speed increases power consumption and, consequently, power dissipation while die size remains the same. Therefore, the power/density ratio will keep increasing to the point where no practical technique can feasibly dissipate the generated heat.

Reliability: Performance, heat and power consumption are not the only concerns. Reliability of Intra-chip communication is also affected by physical constraints. Transistor scaling shortens wire distances, thus improving performance, but it also implies thinning of those wires.

As wires become narrower, they also become taller in order to reduce the resistance per unit length. Media resistance limits the speed of electrons within. Tall wires within close distance vary dependent timing characteristics at best and produce data corruption at worst. In short, thinner wires increase delays and harm reliability.

Furthermore, as explained earlier, the same radiation fluxes that in the past had no effect on electronics are now able to induce faults that affect the logic value of current transistors with lower critical charge.

For these reasons, it is commonly believed by the research community that the classic Hardware/Software uniprocessor model has reached the power/performance-wall (Asanovic et al. 2006; Hill 2001).

Parallelization: The microprocessor industry approach is based on using the billions of transistors now available on a die to replicate the off-the-shelf core design multiple times and increase the size of caches.

Nevertheless, effective programming of multi-core is not trivial and introduces multiple challenges (Geer 2007; Goth 2009; Pankratius et al. 2009). As an attempt to overcome the power wall, research on computer science has reincarnated parallel computing. Parallel computing and parallel programming are not new; they have been a mainstay in high-performance since the early 50s (Hill and Rajwar 2001).

Complexity: The semiconductor industry, driven by economic reasons and time-to-market needs, has introduced unjustified complexity in microprocessor designs. An efficient and logical design could have achieved better results in the long-term. Instead, a brute force approach, increasing frequency and adding deeper pipelines and cache levels has been followed.

Software and hardware inefficiency: In terms of software, modular programming (Turski and Wasserman 1978; Wirth 1983) and later object oriented programming (Wirth 1988) were introduced to maximize performance and effectiveness of the human agent in the programming process.

To maximize performance of HW/SSW/ASW, several approaches of parallelism using distributed, data-flow and cluster architectures were introduced in the late 50s. The Flynn diagram (Flynn 1972) is still in use: SIMD.

SIMD (Single Instruction Multiple Data), MIMD (Multiple Instruction Multiple Data) and MISD (Multiple Instruction Single Data) are very well known architectures, each with their own benefits and drawbacks. In the early 80s the VLIW (Very Long Instruction Word) (Fisher 1983) approach was also introduced. But since then, there were no introduction of significantly new architecture.

The size of the market is in the order of billions of controllers produced every year, which exemplifies the scale of dependency of human society on computer technologies. To make the next step in the design of special systems for safety critical applications we should analyze what is applicable from the well-developed theory and design of fault tolerant systems since early 70s, in particular their reliability and resilience to electromagnetic impulses.

In turn, the success of future computer systems for safety critical applications will depend on trading-off performance, reliability and power consumption.

The combination of these two statements forms a framework for this research. At first, we should analyze the technological achievements of modern electronics in terms of performance. Finally, we should find ways to improve the efficiency of current embedded systems in terms of performance, reliability and power consumption.

1.2 Scope and Contribution

The main contributions of this work relate to the reliability of embedded systems with regards to permanent and transient hardware faults induced by radiation as the source of errors. However, the proposed techniques are also efficient to mitigate the effect of faults induced by other means. Note that software faults as the source of errors are out of the scope of this book. This section briefly explains these contributions.

The main goal of this research was to find efficient techniques and original mechanisms to contribute to improve the reliability, performance, and energy use of real time systems in safety-critical applications. This included the design, development and analysis of a fault-tolerant reconfigurable architecture in presence of radiation-induced faults. Such architecture will be further used as a core element for reconfigurable computers with key requirements of reliability, power awareness, performance and scalability.

This work is an attempt to overcome the known drawbacks of modern RTS. The outcomes of such work can be summarized as follows:

1. The traditional Reliability, Fault Tolerance and Dependability concepts and definitions do not take into account the transient nature of some of the faults induced by radiation. The result is a new concept of resilience that takes into account the changing nature of environment and the different FT contexts;
2. We provided a systematic examination of the physical effects that lead to faults induced by radiation and the error process. The result is a comprehensive taxonomy of radiation-induced effects in modern microprocessor technologies;
3. We developed fault model that contains an extensive taxonomy of faults that can assist in the serviceability and coverage attributes of fault-tolerant and resilient system designs;
4. We introduced a novel combination of structural hardware elements at the active, passive and interfacing zones. In combination with system software, these hardware elements can improve the resilience of a system with a better compromise in silicon area, reliability, power and performance that known fault tolerant systems;
5. We designed and implemented a hardware prototype as a proof-of-concept with core elements introduced in the previous statement;
6. We developed a framework and testing scheme for the testing and debugging of the hardware prototype. As a part of the framework, we implemented an assembler for the hardware prototype together with a disassembler and simulator tool.

Our hardware design work, development of simulation has benefited from collaboration and can be considered as joint international research to make so-called Evolving Reconfigurable Architecture (ERA) with ETHZ Dr T. Kaegi, Dr E. Zoueff, Prof L. Blaeser published in Schagaev et al. (2010).

We proposed a framework and testing scheme for the testing and debugging of the hardware prototype. As part of the framework, we implemented an assembler for the hardware prototype together with a disassembler and simulator tool.

This book is further development of our research effort initiated internationally so-called Evolving Reconfigurable Architecture (ERA) (Schagaev et al. 2010). Theoretical development and hardware testing of ERA will provide the hardware prototype platform for testing hardware reconfigurability.

Chapter contributors are mentioned in each chapter.

References

Adams JH, Gelman A (1984) The effects of solar flares on single event upset rates. IEEE Trans Nucl Sci 31:1212–1216. https://doi.org/10.1109/TNS.1984.4333485

Adams JH, Silberberg R, Tsao CH (1982) Cosmic ray effects on microelectronics. IEEE Trans Nucl Sci 29:169–172. https://doi.org/10.1109/TNS.1982.4335821

Antola A, Erényi I, Scarabottolo N (1986) Transient fault management in systems based on the AMD 2900 microprocessors. Microprocess Microprogram 17:205–217

Asanovic K, Bodik R, Catanzaro B, Gebis J, Husbands P, Keutzer K, Patterson D, Plishker W, Shalf J, Williams S, Yelick K (2006) The landscape of parallel computing research: a view from Berkeley. Technical report No. UCB/EECS-2006-183

Baumann R (2002) The impact of technology scaling on soft error rate performance and limits to the efficacy of error correction. In: International electron devices meeting, 2002. IEDM'02 digest, pp 329–332

Baumann R (2005) Soft errors in advanced computer systems. IEEE des Test Comput 22:258–266

Binder D, Smith EC, Holman AB (1975) Satellite anomalies from galactic cosmic rays. IEEE Trans Nucl Sci 22:2675–2680. https://doi.org/10.1109/TNS.1975.4328188

Blake JB, Mandel R (1986) On-orbit observations of single event upset in Harris HM-6508 1K RAMS. IEEE Trans Nucl Sci 33:1616–1619. https://doi.org/10.1109/TNS.1986.4334651

Blandford JT, Waskiewicz AE, Pickel JC (1984) Cosmic ray induced permanent damage in MNOS EAROMs. IEEE Trans Nucl Sci 31:1568–1570

Fieseler PD, Ardalan SM, Frederickson AR (2002) The radiation effects on Galileo spacecraft systems at Jupiter. IEEE Trans Nucl Sci 49(6):2739–2758

Fisher JA (1983) Very long instruction word architectures and the ELI-512. In: Proceedings of the 10th annual international symposium on computer architecture. ACM, Stockholm, Sweden, pp 140–150

Flynn M (1972) Some computer organizations and their effectiveness. IEEE Trans Comput C 21:948

Geer D (2007) For programmers, multicore chips mean multiple challenges. Computer 40:17–19

Goth G (2009) Entering a parallel universe. Commun ACM 52:15

Hauge PS, Ziegler JF, Srinivasan GR (1996) Special issue: terrestrial cosmic rays and soft errors [WWW document]. IBM J Res Dev. URL http://portal.acm.org/citation.cfm?id=226354. Accessed 7 Jan 2010

Hill MD, Rajwar R (2001) The rise and fall of multiprocessor papers. In: The international symposium on computer architecture (ISCA) [WWW document]. URL http://pages.cs.wisc.edu/~markhill/mp2001.html. Accessed 4 Aug 2010

ITRS (2011) International technology roadmap for semiconductors

Johnston A, Swift G (1997) TOPEX/Poseidon radiation issues: displacement damage in optocouplers

Kish LB (2002) End of Moore's law: thermal (noise) death of integration in micro and nano electronics. Phys Lett A 305:144–149

Mavis DG (2002) Single event transient phenomena—challenges and solutions. Presented at the microelectronics reliability and qualification workshop

Mavis DG, Eaton PH (2002) Soft error rate mitigation techniques for modern microcircuits. In: 2002 40th annual reliability physics symposium proceedings, pp 216–225

Nicolaidis M (ed) (2010) Soft errors in modern electronic systems, 1st edn. Springer

Normand E, Wert JL, Quinn H, Fairbanks TD, Michalak S, Grider G, Iwanchuk P, Morrison J, Wender S, Johnson S (2010) First record of single-event upset on ground, cray-1 computer at Los Alamos in 1976. IEEE Trans Nucl Sci 57:3114–3120

Pankratius V, Jannesari A, Tichy WF (2009) Parallelizing Bzip2: a case study in multicore software engineering. IEEE Softw 26:70–77

Poivey C, Gee G, LaBel KA, Barth JL (2004) In-flight observations of long-term single event effect (SEE) performance on X-ray timing explorer (XTE) solid-state recorders (SSRs) [SRAM]. In:

2004 IEEE radiation effects data workshop (IEEE Cat. No. 04TH8774), July 2004. IEEE, pp 54–57
Pritchard BE, Swift GM, Johnston AH (2002) Radiation effects predicted, observed, and compared for spacecraft systems. In: IEEE radiation effects data workshop, Phoenix, AZ, USA, pp 7–13. https://doi.org/10.1109/REDW.2002.1045525
Schagaev I, Kaegi T, Gutknecht J (2010) ERA: evolving reconfigurable architecture. In: Proceedings of 11th ACIS. Presented at the international conference on software engineering artificial intelligence, networking and parallel/distributed computing, London
Seifert N, Zhu X, Massengill LW (2002) Impact of scaling on soft-error rates in commercial microprocessors. IEEE Trans Nucl Sci 49:3100–3106
Shivakumar P, Kistler M, Keckler SW, Burger D, Alvisi L (2002) Modeling the effect of technology trends on the soft error rate of combinational logic. In: Proceedings of the international conference on dependable systems and networks (DSN 2002), pp 389–398
Swift G, Guertin S (2000) In-flight observations of multiple-bit upset in DRAMs. In: Proceedings of the 37th nuclear and space radiation effects conference, vol 47, no 6, Reno, Nevada, pp 2386–2391
Taber AH, Normand E (1992) Investigation and characterization of SEU effects and hardening strategies in avionics. IBM report No. 92-L75-020-2. Defense Technical Information Center
Taber A, Normand E (1993) Single event upset in avionics. IEEE Trans Nucl Sci 40:120–126
Turski WM, Wasserman AI (1978) Computer programming methodology. SIGSOFT Softw Eng Notes 3:20–21
Waskiewicz AE, Groninger JW, Strahan VH, Long DM (1986) Burnout of power MOS transistors with heavy ions of Californium-252. IEEE Trans Nucl Sci 33:1710–1713. https://doi.org/10.1109/TNS.1986.4334670
Wirth N (1983) Programming in Modula-2. Springer-Verlag TELOS
Wirth N (1988) The programming language Oberon. Softw Pract Exp 18:671–690
Ziegler JF (1996) Terrestrial cosmic rays [WWW document]. IBM J Res Dev. URL https://doi.org/10.1147/rd.401.0019

Chapter 2
Resilience

Abstract This chapter is a background of necessary concepts in the field of fault tolerance and resilience. At first, we introduce the system failure lifecycle and describe the main threats for the systems, justifying the need in providing a new property. Then, the concept of resilience and its attributes and measures are reviewed. We present the vicious cycle, a view of the performance and reliability problems that the microprocessor industry is currently facing. The classic theory of reliability is used and to demonstrate how the hardware components of an embedded system should be done to make system resilient to hardware faults. Extracting essential reliability theory we used to analyze how reliability of a system depends on the topology and features of components. Other attributes of resilience including safety, performability, integrity, maintainability and availability are also reviewed. Finally, we extend the definition of resilience and apply it to the field of safety critical computing.

Keywords System failure · Life cycle · Resilience · Reliability · Performance · Fault · Error · Failure

2.1 System Failure Lifecycle

Correct service (Laprie et al. 1995), also named *proper service* (Laprie and Avizienis 1986), is delivered by a system when the service implements the function as specified. The fundamental threats to the correct service and to the resilience of safety critical systems are faults, errors and failures that, in turn, can cause catastrophic failures. Among these four terms there is a cause-effect relationship.

A *failure*, service failure or system failure is an event that takes place when the delivered service deviates from *proper service*. Hence, a service failure implies a transition of the system from proper service to an improper *service*, not implementing the functions as specified by the functional specification of the system. The downtime or period of delivery of improper service is also referred to as service *outage*. The transition from improper service to proper service is called *service restoration*, *service recovery* or *repair*.

© The Author(s), under exclusive license to Springer Nature Switzerland AG 2026

I. Schagaev, *Resilience of Computer Systems*,
https://doi.org/10.1007/978-3-032-00938-8_2

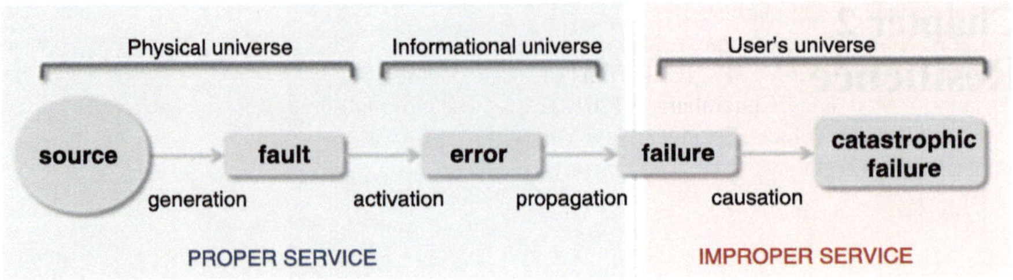

Fig. 2.1 System failure lifecycle within a three-universe model

Since a service is a sequence of the external states of a system, a service failure takes place when one or more of its external states deviate(s) from the correct service state. These deviations are errors. An error is a part of the system state that is liable to lead to a subsequent failure. The hypothesized or adjudged cause of such error is a fault.

A *fault* is a weakness, blemish or shortcoming of a particular hardware component or unit. An error is the manifestation of a fault, a deviation from accuracy or correctness. Finally, if the error leads to one of the system's functions being performed incorrectly then a failure has occurred.

Figure 2.1 graphically describes the well-known lifecycle of system failure within a three universal models (Johnson 1989) adapted from the four universe model originally developed by Avizienis (1982). In the first universe, the physical one, faults are generated due to various sources. Faults can activate errors within the second universe, the informational one.

Errors take place when some information units become incorrect. In turn, errors could propagate the user universe and lead to a failure. It is in this final universe, where the user can witness the effects of faults and errors in the form of failures. One or more failures could potentially cause a catastrophic failure in the case of safety critical systems.

The arrows between the entities in Fig. 2.1 correspond to latencies. *Fault latency* (activation latency in Fig. 2.1) is the time length between the occurrence of a physical fault and the appearance of an error. Likewise, *error latency* is the length of the propagation time that takes place between the activation of the error and the manifestation of the failure.

The term fault and failure is sometimes unclear in reliability literature. In this work, the term fault is sometimes equivalent to failure. For instance, a *system fault* can be the same as a *component failure*. Figure 2.2 shows the fault-failure transition between a subsystem and a global subsystem.

The fault-failure cycle can be applied at different levels of abstraction within a system; consider a transistor as a subsystem that is part of a more global system (e.g. memory cell): the occurrence of incorrect functionality of the transistor during normal operation (e.g. the effects of ageing and stress) is a subsystem failure of such component but may lead to, for instance, a logic fault (global system fault).

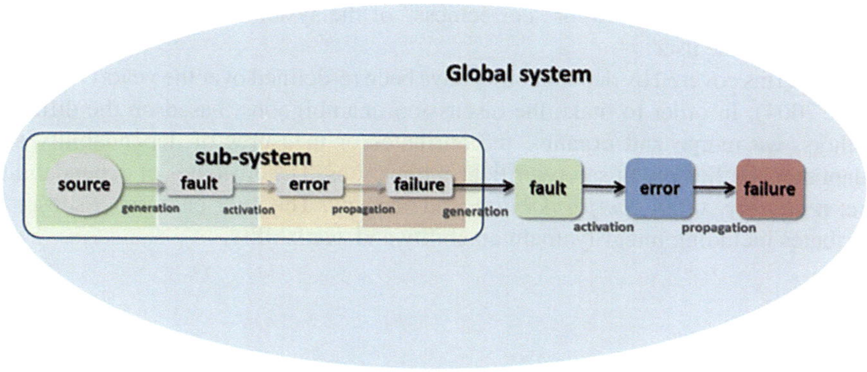

Fig. 2.2 Failure-fault transition between different levels of a system

This logic fault will remain dormant unless is activated, producing an error, which is likely to propagate and create other errors. If the correct service of that global system is affected, a global system failure occurs. The same subsystem-system transition can take place between the memory cell, the memory circuit that the cell is part of, the microprocessor system that can be part of a multiprocessor, etc.

2.2 Attributes and Measures of Resilience

The word *resilience* from the Latin etymology *resilire* (to jump back, or to rebound) is literally the tendency, ability, act or action of springing back, and thus the ability of a body to recover its normal shape and size after being pushed or pulled out of shape. That is, the ability to recover to normality after a disturbance, shock or deviation from the intended state and go back to a pre-existing or acceptable or desirable, state.

The meaning of resilience is somewhat different between authors. The term *Resilient* has been traditionally used essentially as a synonym of fault-tolerant (Laprie 2008). Before we discuss fault tolerance as a concept and introduce resilience, several other terms need to be defined.

One of them is *Dependability*, which is an integrative concept that encompasses many other quantitative and qualitative attributes. Laprie (Laprie et al. 1992) defines *dependability* as the "trustworthiness of a computer system such that reliance can be justifiably placed on the service that it delivers".

Dependability is the ability to deliver a service that can justifiably be trusted. Laprie defines the service delivered by the system as its behavior as it is perceptible by is user(s); a user is another system physical or human which interacts with the former. Such service is classified as "proper" or "correct" if it is delivered as specified; otherwise it is considered as "improper" or "incorrect" (Laprie and Avizienis 1986).

Again, the "properness" or "correctness" of the system service depends on the viewpoint of the user.

The terms covered by dependability have been re-defined over the years (Avizienis et al. 2004). In order to make the discussion unambiguous, based on the different authors, we merge and organise the attributes or measures of dependability and adapt them to the field of safety-critical applications. The attributes of dependability are: *reliability*, *safety*, *performability*, and *security*. The later encloses a subset of attributes including integrity, maintainability and availability.

2.3 Reliability

The reliability measure is most often used to characterize systems in which failures are unacceptable; therefore, it is suitable to the field of safety critical systems.

Figure 2.3 shows a non-repairable system with two possible states: a fully functional start state (up) and a failed state (down), involving loss of functionality, which can be reached after a transition due to failure.

There is no disagreement about the need for reliable systems but some vague notion of reliability is not enough in safety-critical engineering. Reliability can be defined as follows:

Reliability R(t) is the probability that a system or component will perform its intended function (without failure) over the entire interval [0, t] under specified environmental and operating conditions.

2.3.1 Performance and Reliability

2.3.1.1 Power-Reliability Wall

Since the invention of the integrated circuit in 1958, each generation of semiconductor technology has exponentially decreased the transistor price and exponentially increased the transistor density per chip (Hutcheson 2009).

This technological shrink model has led to the impressive level of technology and hardware element density recently achieved (Nair 2002) with processor frequencies

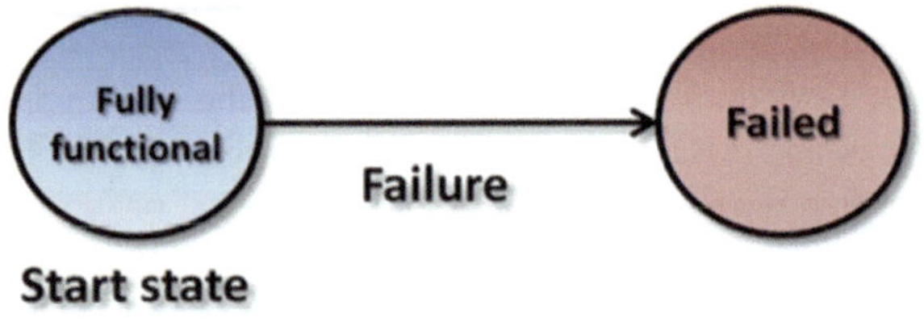

Fig. 2.3 A non-repairable system with two states

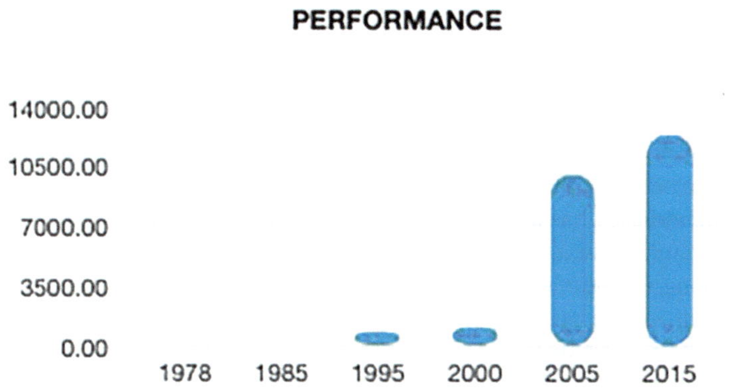

Fig. 2.4 Growth of performance since the mid-80s

reaching up to 4.7 GHz "Power 6 Specs: IBM Power6 Microprocessor and IBM System p 570," 2007. The higher number of transistors and the kilometres of wire operating at higher frequencies lead to several W/cm^2 on modern chips leading the peak energy consumption well over 140 W. Most of that energy becomes heat, rising operating temperatures.

Heat and power consumption are not the only concerns of modern electronics. Reliability and performance of intra-chip communication is also affected by physical constraints. Transistor scaling shortens wire distances, improving performance, but it also implies thinning of those wires. The resistance and capacitance of the media limit the speed at which electrons can flow. Thinner wires increase delays and harm reliability (Shivakumar et al. 2002). For these reasons, the research community believes that the classic HW/SW single processor model has reached the *power-reliability wall* (Asanovic et al. 2006).

Moreover, the cost of the manufacturing process technologies needed to continue Moore's law is increasing exponentially. Such cost is doubling every four years, which makes smaller nanometre technologies and the continuation of this law not only a technical challenge but an economic one.

Evidence of this phenomenon is the chart shown in Fig. 2.4 that plots performance gap between the processor and memory of the VAX 11/780 measured by the *SPECint benchmarks.*[1] This data are taken from Patterson and Hennesy (2012). Subsequent to the mid-80s, processor performance growth averaged about 52% per year. Since 2002, uniprocessor performance has slowed down to about 20% per year reaching the power-reliability wall in 2006. On the other hand, memory has averaged a constant performance increase of 9%.

Using Moore's law as a metric of progress has become misleading, as improvements in transistor density no longer translate into performance and energy efficiency. Starting around the 65 nm technologies, transistor scaling no longer delivers

[1] SPECint benchmarks are a set of benchmarks design to test the integer processing performance of modern CPU (http://www.spec.org/).

the performance and energy gains that drove the semiconductor growth during the past decades (Dreslinski et al. 2010). In our recent book (Schagaev and Gutknecht 2024), Chap. 15 shows that book grown in performance, say, 30% will need at least duplication of hardware.

The research community and the Industry believe that parallelism is the answer to overcome the performance wall, however, with different approaches on how to implement it. The industry has attempted to react by escalating the number of processors introducing multi-core architectures and parallelism.

Multiplying the number of big, complex and power demanding existing cores, which are part of the problem, does not adequately solve any of the performance, reliability and power awareness concerns (Asanovic et al. 2006). Replicating the essence of the problem will hardly achieve a solution to it.

2.3.1.2 Reliability with Vicious Cycle

The semiconductor industry, driven by economic reasons and time-to-market rush has introduced too much complexity in microprocessor designs. Figure 2.5 shows our interpretation of the reliability problem in current computing.

An efficient, weighted, deeper and logical design could bring way better results in short and long terms. Instead, a brute force approach, increasing frequency, longer

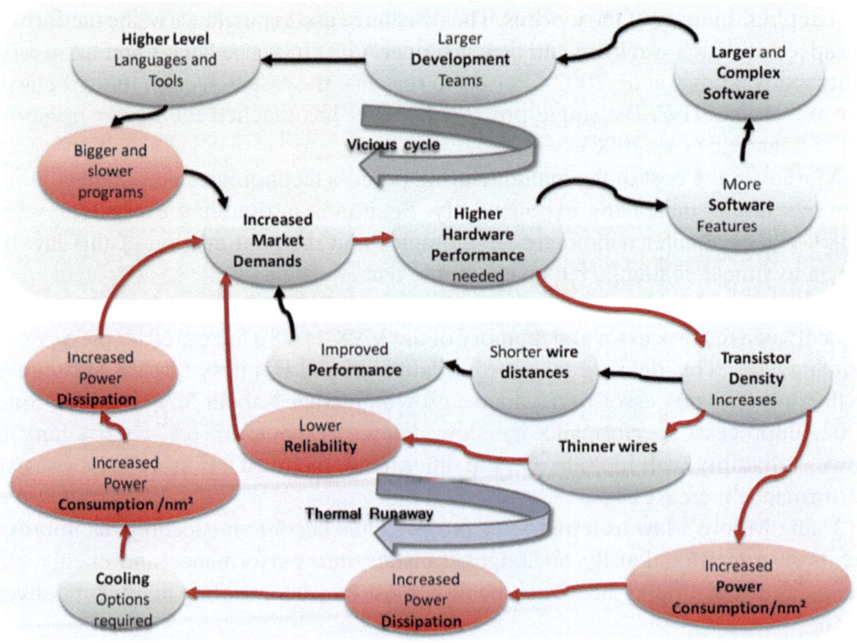

Fig. 2.5 The vicious cycle and the evolution of computing systems, 1950–2005

pipelines unjustified multithreading and cache levels provide slight performance at a high cost of chip floor plan, see Asanovic et al. (2006).

Increased processor performance allows software companies to develop larger and feature-richer software, which involves larger development teams. Consequently, developers need higher-level languages and abstractions, which are less efficient and generate slower programs. As a result, faster processors are needed, reinforcing this vicious cycle (Fig. 2.5) and generating detrimental results. Under this cycle, existing programs would run faster on the latest generation of microprocessors.

Since 2005–2006, no considerable increase in functional hardware performance has occurred. Existing programs need to be redeveloped to take advantage of the new multi-core. Consequently, the vicious cycle does not apply anymore. The power wall has dramatically slowed down the evolution of microprocessors in terms of performance.

Clearly, technological developments have not been supported by a logical evolution. There is an increasing need for unified hardware and software technologies. Development of a new computing paradigm and its implementation through the whole cycle of hardware, software and application design, development and prototyping is required.

2.3.2 Reliability and Unreliability Functions

Let's suppose a system with N identical components. We define $S(t)$ as the number of surviving components at time t and $Q(t)$ as the number of failed components up to time t. Therefore:

Equation 2.1 Surviving and failed components at time t

$$S(t) + Q(t) = N \tag{2.1}$$

The *reliability*$R(t)$ is the proportion of components that continue to perform without failure after being used for a period of time t. That is *the probability* of survival of the components, given by:

Equation 2.2 Probability of survival of components at time t

$$R(t) = \frac{S(t)}{N} \tag{2.2}$$

Unreliability, or *Cumulative failure distribution function* is generally referred to as the *probability of failure*. More specifically, *unreliability* $F(t)$ is the *conditional probability* that the system begins to perform incorrectly during the interval $[t_0, + t]$, given that the system was performing correctly at time t_0:

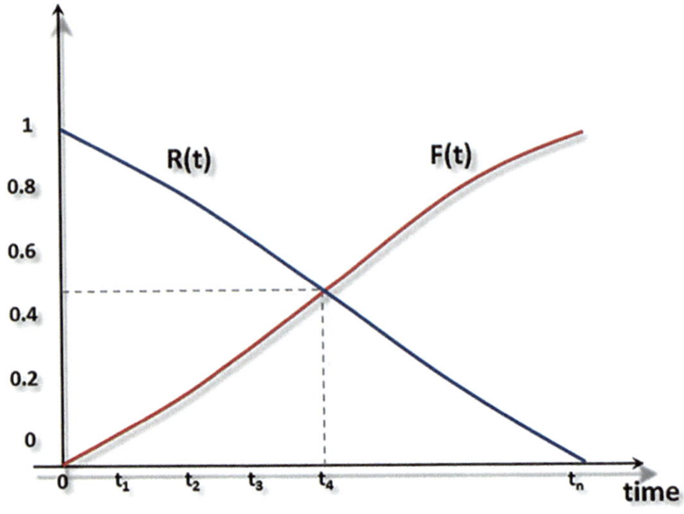

Fig. 2.6 Reliability $R(t)$ and probability of failure $F(t)$ functions over time t

Equation 2.3 Probability of failure of components at time t

$$F(t) = \frac{Q(t)}{N} \tag{2.3}$$

Based on Eq. (2.1):

Equation 2.4 Reliability and probability of failure of components at time t

$$\begin{aligned} R(t) + F(t) &= 1 \\ F(t) &= 1 - R(t) \end{aligned} \tag{2.4}$$

In turn, Fig. 2.6 shows a graph of the reliability and probabilities of failure over time with a constant failure rate. $R(t)$ is a monotonically decreasing function that has an initial value 1 whereas $F(t)$, starting at 0, increases monotonically. The sum of $F(t)$ and $R(t)$ at any given time is 1.

2.3.3 Probability Density Function

The derivative of $F(t)$ is a probability distribution function (PDF) that defines the probability of failures per unit time $F(t)$ of a particular component that has been used for a period of time t. Based on this definition, the probability density function is described as:

Equation 2.5 Probability density function as a function of unreliability

$$f(t) = \frac{\mathrm{d}F(t)}{\mathrm{d}t} \tag{2.5}$$

Using Eq. (2.4):

Equation 2.6 Probability density function as a function of reliability

$$f(t) = \frac{\mathrm{d}[1 - R(t)]}{\mathrm{d}t} = -\frac{\mathrm{d}R(t)}{\mathrm{d}t} \tag{2.6}$$

Thus, the probability of a failure during the time range [0, t] is:

Equation 2.7 Probability of failure during time range [0, t]

$$F(t) = \int_0^t f(t)\mathrm{d}t \tag{2.7}$$

Again, using Eq. (2.4) for Reliability over range [0, t]:

Equation 2.8 Reliability during time range [0, t]

$$R(t) = 1 - F(t) = 1 - \int_0^t f(t)\mathrm{d}t = \int_t^\infty f(t)\mathrm{d}t \tag{2.8}$$

Schematically, Fig. 2.7 presents Reliability, Unreliability (red and blue areas, respectively) and PDF.

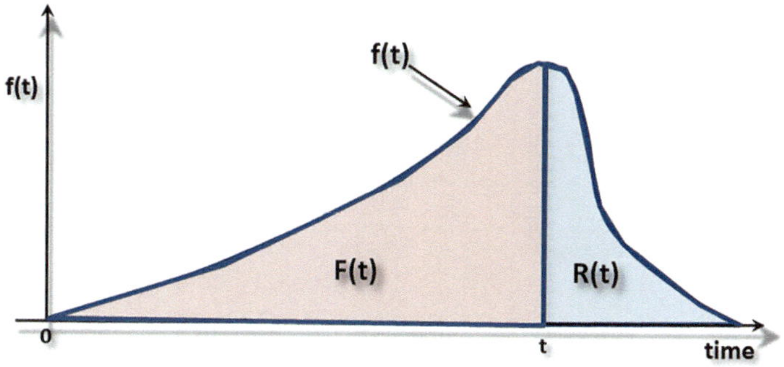

Fig. 2.7 Representation of reliability, unreliability and the probability density function

2.3.4 Failure Rate Function

The failure rate function $\lambda(t)$ also known as *momentary failure rate* or *hazard function* describes the number of failures per unit of time versus the number of components still operating at a time surviving components:

Equation 2.9 Failure rate as failures versus components at time t

$$\lambda(t) = \frac{1}{S(t)} \frac{-\mathrm{d}Q(t)}{\mathrm{d}t} \tag{2.9}$$

Using Eqs. (2.2) and (2.3):

Equation 2.10 Failure rate as a function of reliability and probability density

$$\begin{aligned} \lambda(t) &= \frac{1}{N\,R(t)} \frac{N\,\mathrm{d}F(t)}{\mathrm{d}t} \\ \lambda(t) &= \frac{1}{R(t)} \frac{\mathrm{d}F(t)}{\mathrm{d}t} = \frac{f(t)}{R(t)} \end{aligned} \tag{2.10}$$

The failure rate function is very accurate to express the reliability of semiconductor components for long periods.

However, calculating the failure rate at a specific point of time within a short period is impractical. Consequently, *average failure rate*, with longer time periods, is preferred:

Equation 2.11 Average failure rate

$$\text{Average failure rate} = \frac{\text{Total failures during a period}}{\text{total operating time within a period}} \tag{2.11}$$

The values of average failure rate can be expressed by % or ppm.[2] However, FIT[3] it is more widely used as a unit for reliability.

[2] ppm is the abbreviation of "parts per million". One ppm means 1 faulty component out of 1,000,000 components. Hence, an average failure rate of 10 ppm means that there are 100 faulty components out of 1,000,000, or 1 component out of 100,000.

[3] FIT is a unit widely used to express failure rate. One FIT equals to one failure per billion (10^9) hours (one failure in about 114,155 years), or 1 ppm/1000 h.

2.3.5 Cumulative Hazard Function

Using Eq. (2.4), failure rate as a function of reliability can be presented as:

Equation 2.12 Failure rate function as a function of reliability

$$\lambda(t) = \frac{-1}{R(t)} \frac{\mathrm{d}R(t)}{\mathrm{d}t} \tag{2.12}$$

The limits of integration can be obtained as follows:

- at time $t = 0$, $R(t) = 1$
- at time t by definition the reliability is $R(t)$.

Given the assumption of a constant failure rate λ of a component (typically in per million hours or *FIT*):

Equation 2.13 Reliability at time t with constant failure rate λ

$$\begin{aligned} \lambda t &= - \log R(t) \\ -\lambda t &= - \log R(t) \\ R(t) &= \mathrm{e}^{-\lambda t} \end{aligned} \tag{2.13}$$

2.3.6 Bathtub Curve of Failure Rates

A value can be assigned to the reliability of a system. For instance, a system may have 97% reliability over a two year mission, subject to a maximum vibration $V_{\max}$, a humidity range $[H_{\min}, H_{\max}]$ and temperature range [15 °C, + 30 °C]. Although the above definition is generally accepted, it is not a complete definition from the starting to the end time of a safety-critical system's life. System reliability will be different for different time periods. Therefore, more factors need to be considered. For a correct service delivery in a specific period, the system must be operating properly at the beginning of the observation period.

The age of the system is one of the factors that should be taken into account. The above definition does not differentiate between:

1. new system,
2. a system that has been operational for a substantial amount of time and whose faults have already been corrected, and
3. an old system with a long operational history and wear out issues.

Reliability distributions with decreasing, constant and increasing failure rate as a function of time are illustrated in Fig. 2.8 during period (a), (b) and (c), respectively.

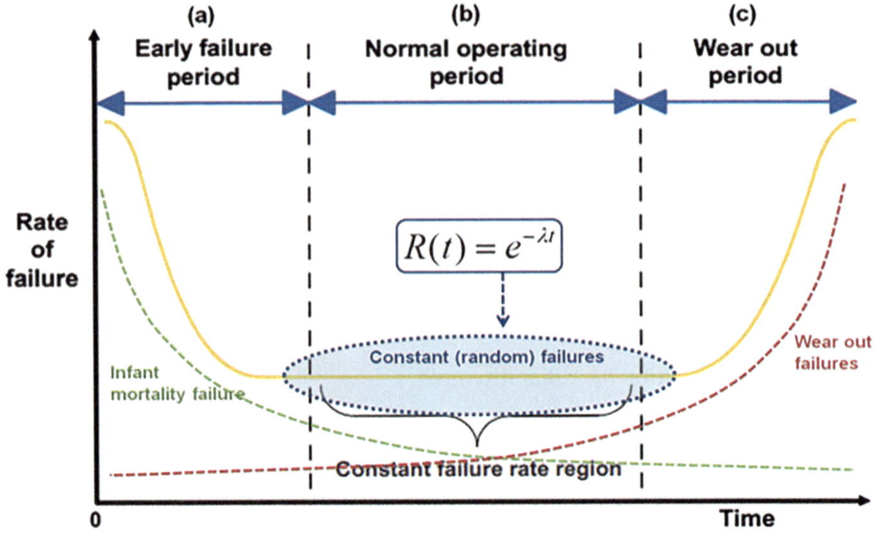

Fig. 2.8 A bathtube curve of failure rates. During normal operation period the failure rate λ is constant and faults are independent

The assumption made is that faults are independent and that the failure rate (λ) is constant. The system failure rate is dependent on the system's lifetime constituting a function with a bathtub shape and three distinctive areas or periods: an early failure period (a), a normal operating period (b) and a wear out period (c). For failure rates higher than the constant failure rate (λ), the chance of system failure becomes higher.

For a new system (case a) there is an *early failure* or *infant period* with a decreasing but high failure rate due to latent manufacturing defects that escape the initial testing of the product. As the products get into operation, these defects surface quickly when the devices are stressed.

Once the infant failures are eliminated, this high failure rate rapidly decreases to an almost constant value during the *normal + operating or grace period* (case b). This long period represents the useful life of the system where failures occasionally occur due to the sporadic breakdown of weak components. It is highly desirable that this period of low failure rate and high reliability dominates the product's lifetime.

During the *wear out* or *breakdown period* (case c) the reverse situation takes place. As the system gets older, the failure rate increases sharply due to age-related wear out. Note that many devices that form part of the same system will initiate this phase roughly at the same time. This could create an avalanche effect that could critically decrease the overall reliability of the system.

After analysing the bathtub curve and the three periods of operation involved, it is clear that the previous equations of reliability only suit the normal operating period with a constant failure rate. This curve represents very well hardware reliability due to ageing and degradation but it is not suitable to software, especially in the case of versioning and upgrades. The silicon failure mechanisms will be further studied in Chap. 4.

2.3.7 Mean Time Between Failures (MTBF)

Instead of a monotonic function of time reliability can also be expressed as a single numeric index. Mean time between failures (MTBF) is the average time that the system will run between failures. This measure is convenient to compare the reliability of different repairable systems. MTBF can be estimated by averaging the time between failures, including any additional time required to repair the system and place it back to a functional state. Being $f(t)$ the probability of failure per unit time, MTBF can be described by:

Equation 2.14 Probability of failure per unit time (MTBF)

$$\mathrm{MTBF} = \int_0^{\infty} t f(t)\mathrm{d}t \tag{2.14}$$

Using Eq. (2.6) MTBF then:

$$\mathrm{MTBF} = -\int_0^{\infty} t \frac{\mathrm{d}R(t)}{\mathrm{d}t} \tag{2.15}$$

Integrating the above equations by parts we obtain:

$$\mathrm{MTBF} = -[tR(t)]_0^{\infty} \int_0^{\infty} R(t)\mathrm{d}t \tag{2.16}$$

For $t = 0$, $R(t) = 0$, hence $t \times R(t) = 0$. As t increases from 0, $R(t)$ decreases. As t tends to ∞, $t \times R(t)$ tends to zero. Therefore, the first term of the previous equation is zero. For any kind of failure distribution with a failure rate λ as a function of time, the general expression for MTBF can be described as:

Equation 2.17 General expression of MTBF

$$\mathrm{MTBF} = \int_0^{\infty} R(t)\mathrm{d}t \tag{2.17}$$

The higher the MTBF is, the higher is the reliability of the system or component. Moreover, for failure distributions independent of time with a constant rate, MTBF is given by:

Equation 2.18 MTBF for time independent failure distributions with constant rate of failures

$$\mathrm{MTBF} = \int_0^\infty \mathrm{e}^{-\lambda t} \mathrm{d}t \tag{2.18}$$

Equation 2.19 MTBF for time independent failure distributions with constant rate of failures

$$\mathrm{MTBF} = \frac{1}{\lambda}\left[\mathrm{e}^{-\lambda t} \mathrm{d}t\right]_0^\infty = \frac{1}{b} \tag{2.19}$$

Hence, MTBF of a system is reciprocal to its failure rate (given a constant failure rate). MTBF will be expressed in hours if the constant rate is also expressed in hours.

2.3.8 *Mean Time to Failure (MTTF)*

As described above, MTBF is a good measure of reliability for systems that can be repaired. A similar single-parameter indicator of reliability for components that cannot be repaired is the mean time to failure (MTTF). MTTF is the average time until the first system's failure.

Results of life testing can be used to calculate MTTF by testing a set of N identical units until all of them have failed with the time to the first failure of the individual units identified as $t_1, t_2, t_3, \ldots, t_n$. It can be observed that MTTF is given by:

Equation 2.20 Mean time to failure (MTTF)

$$\mathrm{MTTF} = \frac{1}{\lambda} \sum_{i=1}^{n} t_i \tag{2.20}$$

As before, the failure rate, if independent of time, can be calculated by:

Equation 2.21 Failure rate as the inverse of MTTF

$$\lambda = \frac{1}{\mathrm{MTTF}} \tag{2.21}$$

MTBF and MTTF are sometime used interchangeably. Although, the numerical difference is small in many cases, both measures represent different concepts. MTTF is related to MTBF but does not include the repair time (MTTR or mean time to repair/restore) nor the detection time (MTTD or mean time to detection):

Equation 2.22 MTBF as a function of MTTF, MTTD and MTTR

$$\mathrm{MTBF} = \mathrm{MTTF} + \mathrm{MTTD} + \mathrm{MTTR} \tag{2.22}$$

MTTR is the average time required to repair a system, whereas MTTD is the average time required to detect a failure. In most applications, MTTR and MTTD are just a small fraction of the total MTTF. Therefore, the approximation that MTBF and MTTF are almost equal is sometimes fair.

MTTR and MTTD are difficult to estimate and can be determined by injecting faults into a system, measuring the time required to repair it. Both measures will be further discussed in the availability section.

2.3.9 Reliability Prediction

In the case of design of hardware systems, there are two different known theoretical methods to meet the above mentioned reliability requirements and specifications:

- Fault avoidance: makes use of substantially higher reliability components and substantially higher than expected lifetime. Birolini (2014) introduced a comprehensive theoretical approach based on the application of reliability engineering throughout the system to reach this goal.
- Fault tolerance: deliberately introduces redundancy in the system to achieve continuous operation.

During the last 50 years there have been several attempts (Gnedenko et al. 1999; Koren and Krishna 2007; Kovalenko et al. 1997) to connect probability and reliability. A brief review of the probability theory for the analysis of real objects and their features (fault tolerance) is presented below.

Reliability of systems can be estimated by partitioning those systems into more elemental entities (e.g. subsystems or components) and then by assessing the individual probability theory of these individual entities.

The entities can be interconnected in serial, parallel or both. Therefore reliability models are needed to illustrate the functional relationship among the entities of the system and the way in which a failure of each component would affect the overall reliability of the system.

2.3.9.1 Serial Reliability

In this model, the entities are connected in series. When minimum design and costs are specified in the design requirements of a system, a series system is the usual choice for designers. For the system to be operational, all of the components or subsystems should be operational and work correctly. Serial systems are inherently unreliable since the failure in one of the elements would cause a stoppage of the overall system. The reliability of a system without redundancy may be described with a sequential reliability block diagram (see Fig. 2.9).

In this arrangement the system reliability is the product of its individual component reliabilities, assuming they are organized in serial (cumulative) structure. Note that for this structure, if the reliability of each component is R_i, the total system reliability R_s is given by:

Equation 2.23 System reliability of a serial system

$$R_s(t) = \prod_{i=1}^{n} R_i(t) = \exp\left(-\left(\sum_{j=1}^{n} \lambda_j\right)t\right) \tag{2.23}$$

And the failure rate of the system λ_s is given by:

Equation 2.24 Failure rate of a serial system

$$\lambda_s = \lambda_1 + \lambda_2 + \lambda_3 + \cdots + \lambda_n \tag{2.24}$$

Furthermore, the Mission Time Function $M_T(r)$ gives the time at which system reliability falls below the given threshold level r. The relationship between reliability $R(t)$ and mission time $M_T(r)$ is given by the definitions:

Equation 2.25 Mission time function M_T with threshold level r

$$R[M_T(r)] = r \tag{2.25}$$

Equation 2.26 Mission time function M_T at a given time t

$$M_T[R(t)] = t \tag{2.26}$$

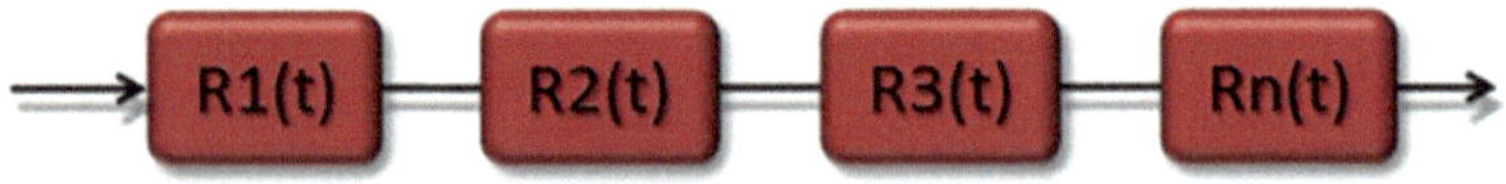

Fig. 2.9 Logic diagram of serial system reliability

If λ is constant then, using Eq. (2.13):

Equation 2.27 Mission time function M_T with constant failure rate

$$t = \frac{-\ln(r)}{\lambda}$$
$$M_T(r) = \frac{-\ln(r)}{\lambda} \tag{2.27}$$

Respectively, for a non-redundant system with n components

Equation 2.28 Mission time function M_T for non-redundant systems with n components

$$M_T(r) = \frac{-\ln(r)}{\sum_{i=1}^{n} \lambda_i} \tag{2.28}$$

The failure rate of a sequential independent element system is equal to the sum of the failure rates of its elements. In the case of a constant failure rate across all elements, the MTTF of the whole system ($MTTF_S$) can be calculated as follows:

Equation 2.29 Mission time to failure of a system with constant failure rate

$$MTTF_S = 1/\lambda_s \tag{2.29}$$

Note that this equation highlights the fact that the reliability of a system is directly impacted (in practice often dominated but not solely determined) by the reliability of its least reliable component.

2.3.9.2 Parallel Systems Reliability: Redundancy and Fault Tolerance

In the previous model, no redundancy was taken into account to calculate the system reliability. A second approach to achieve a required level of reliability is the deliberate introduction of extra components into the system. The sole purpose of introducing this redundancy artificially is to increase reliability. However, there is a price to pay for such improvement in the system's reliability.

This approach assumes a deliberate introduction of redundancy in the system and has been applied since the original work of Von Neumann (1956) and Pierce (1965).

Note that introducing redundancy involves some additional components and complexity and it is therefore imperative that the reliability benefit accruing from the redundancy scheme must far exceed the decrease in reliability due to the actual implementation of the redundancy mechanism itself.

The classic parallel generalization of the redundancy model (Birolini 2014) describes a system of n statistically identical elements in active redundancy, where k element(s) is/are required to perform a function and the remaining n–k are in reserve.

Fig. 2.10 Parallel reliability

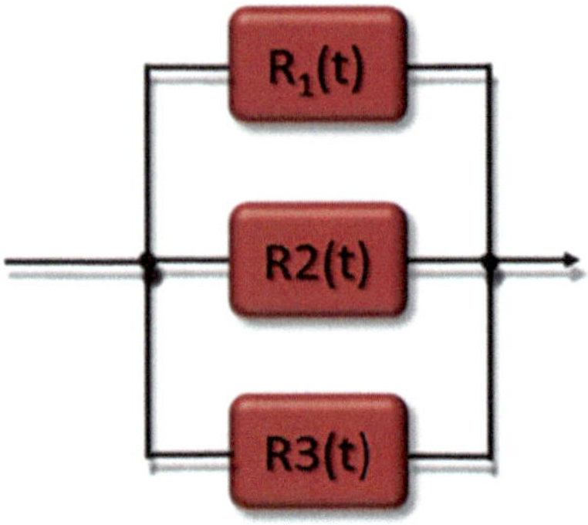

A function of the system is considered successful if during scheduled time k element(s) of the system was/were available. As an example, in the case of a *1-out-of-3* system (Fig. 2.10), its function would be complete if at least one of the elements was known to be working correctly. The second and third elements are redundant and introduced only for reliability purposes when the first unit is known to be faulty.

For the system of Fig. 2.10 the reliability function is as follows:

Equation 2.30 General reliability of a 1-out-of-3 parallel system

$$R(t) = R_1(t) + R_2(t) + R_3(t) - R_1(t)R_2(t)R_3(t) \tag{2.30}$$

Assuming that the elements are identical, work or fail independently of each other and have constant failure rate $R_1(t) = R_2(t) = e^{-\lambda t}$, then by substitution:

Equation 2.31 Reliability of a 1-out-of-3 parallel system with constant failure rate

$$R(t) = 3e^{-\lambda t} - e^{-3\lambda t} \tag{2.31}$$

And mean-time to failure:

Equation 2.32 Mean time to failure of a 1-out-of-3 parallel system with constant failure rate

$$\mathrm{MTTF_s} = \frac{3}{\lambda} - \left(\frac{1}{3}\right)\lambda = \frac{8}{3\lambda} \tag{2.32}$$

Therefore, the apparent working time of the redundant system is increased. In the general case where n redundant elements are introduced as spares to provide successful completion of an element's function with the same assumptions as above, the overall system reliability is given by:

Equation 2.33 General reliability of a 1-out-of-n parallel system with constant failure rate

$$R(t) = 1 - \left(1 - e^{-\lambda t}\right)^n \tag{2.33}$$

In the above equation n is the number of modules, $e^{-\lambda t}$ is the reliability of the original system and it is assumed that:

1. There is a fault-free mechanism to detect and report failure of the active module,
2. There is a fault-free switching mechanism to replace the active module in case of detected failure, and
3. All modules have equal reliability.

Thus, there is no doubt that redundancy even for this classic case could improve reliability of the system considerably. Note that the redundant components do not necessarily need to be identical, but could also correspond to additional hardware with different reliabilities used to detect and treat transient faults.

2.3.9.3 Reliability for Mixed Serial and Parallel System

In practice, systems are usually made of a combination of serial and parallel components. More complex math applies to the reliability of these mixed arrangements. This type of arrangement is frequently used in systems where a specific part is particularly prone to failure. Figure 2.11 depicts an example of M of N system, whose elements may or may not have constant rates, and has a voter that counts for the serial reliability element.

Assuming that only 1 out of N parallel components needs to operate, the reliability of the parallel section of the system is defined by:

Equation 2.34 General reliability of a parallel section of a specific mixed serial/parallel system

$$R(t) = 1 - [(1 - R_1(t))(1 - R_2(t))(1 - R_3(t))] \tag{2.34}$$

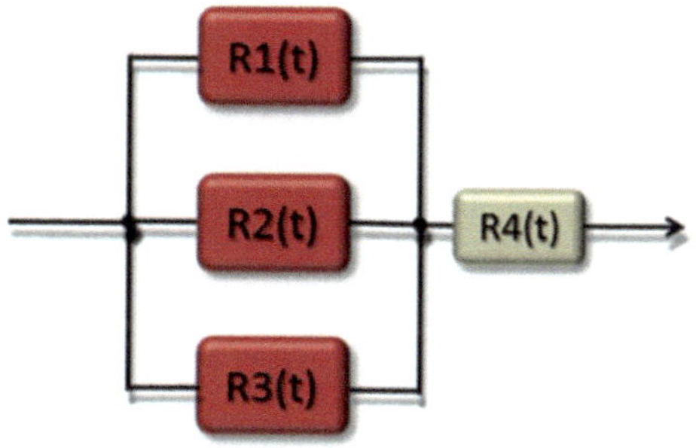

Fig. 2.11 Reliability of a combination of serial/parallel components with a voter

The total reliability of the mixed serial/parallel system shown in Fig. 2.11 is specified by:

Equation 2.35 General reliability of a parallel section of a specific mixed serial/parallel system

$$R(t) = R_{1-3}(t)R_4(t) \tag{2.35}$$

Therefore, a relatively reliable voter would dominate the reliability of a redundant system.

2.4 Safety

In safety-critical systems, safety describes the absence of catastrophic failures for users and the environment when a failure takes place. A system that can be repaired after failure presents a minimum of two states: functional and failed. Some other systems are able to have extra states even under faulty conditions. An example of such system, depicted in Fig. 2.12, has the possibility of transiting to a safe state, in a manner that does not cause any harm.

Safety is a measure of the fail-safe capability of a system and it is defined as the probability that a system will either perform its function correctly or will discontinue its operation in a safe way.

Quantitatively, safety is the probability that the system will not fail in the interval [0, *t*] in such a manner as to cause unacceptable damage to other systems or compromise the safety of any people associated with the system.

The safety function can be described by:

Equation 2.36 Availability as reliability and recoverability

$$S(t) = P_{\text{functional}}(t) + P_{\text{safe-mode}}(t) \tag{2.36}$$

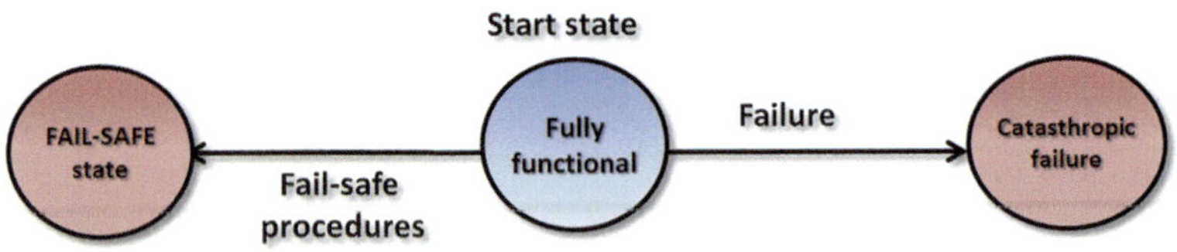

Fig. 2.12 A basic fail-safe system with three states

Safety is directly dependent on "*risk*", as the probability of loss associated to a particular failure. In turn, risk is a function of the probability of failures and their severity on the system. A system can be unreliable, have low availability and yet be safe. A system is safe if it functions correctly or if in case of failure it can remain in a safe state.

2.5 Security

Security is the concurrent existence of three attributes: *integrity*, *maintainability* and *availability*.

2.5.1 *Integrity*

The attribute of Integrity is inward-looking and is related to the capability of a system to protect computational resources and data under severe circumstances. Integrity can be defined as the absence of improper system state alterations. Two types of integrity can be defined:

- System integrity: the ability of a system to detect faults during operation and to inform to a human operator.
- Data integrity: the ability of a system to prevent damage in data and possibly to correct errors that occur as a consequence of faults.

2.5.2 *Maintainability*

Qualitatively, maintainability is referred to as the ease and rapidity in which, following a failure, a repairable system can be restored to a specified operational condition. Quantitatively, maintainability is defined as the probability $M(t)$ that a failed system will restore to a normal operable state specified within a given time frame t.

The restoration process involves the location of the problem, the reparation\recovery of the system bringing it back to a normal operational condition. Maintainability has two main components, serviceability and recoverability that should by carefully analysed in the implementation of self-repairing systems:

Equation 2.37 Maintainability of a system

$$M(t) = f(S(t), \mathrm{RC}(t)) \tag{2.37}$$

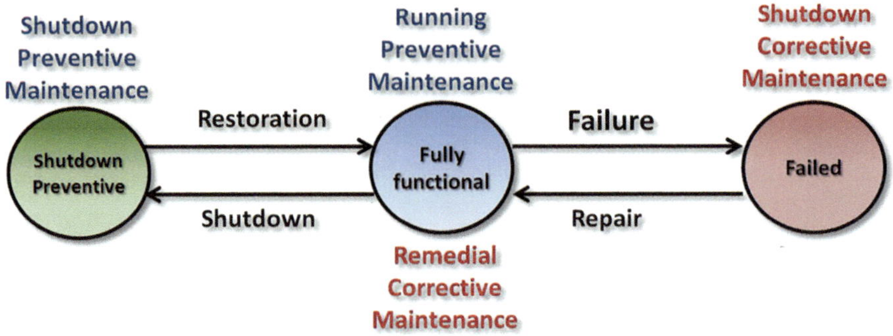

Fig. 2.13 Preventive and corrective maintenance on a three state repairable system

Maintainability characteristics are determined by the system design of maintenance procedures, such as *preventive* (*PM*) and *corrective* maintenance (*CM*) procedures that respectively apply to the serviceability and recoverability components, respectively, and determine the length of repair times (Bodsberg and Hokstad 1995; Dhillon 2006).

PM is the set of activities performed on a system before the occurrence of failure in order to prevent any degradation in its operating condition. *PM* aims to reduce the probability of failure at predetermined intervals or along with prescribed criteria. *CM* is the remedial set of activities performed on a system in order to recover an item to its fully functional condition. CM is usually unplanned that requires urgent attention.

Figure 2.13 shows PM and CM mechanisms on a three-state repairable system. Note that not all maintenance leads to downtime of the three-state system. Whilst *running* PM and *remedial* CM prevent and correct failures during normal operation, *shutdown* PC and CM take place during non-functional states.

Figure 2.13 shows PM and CM mechanisms on a three-state repairable system. Note that not all maintenance leads to downtime of the three-state system. Whilst running PM and remedial CM prevent and correct failures during normal operation, shutdown PC and CM take place during non-functional states.

2.5.2.1 Recoverability

Once the problem has been identified and located by the testability mechanisms, CM can be carried out to complete the necessary repairs. Consider a repairable system with two states: a fully functional and a failed one (as in Fig. 2.3); however, in this case the failed state can be abandoned after successful CM, transiting back to a fully functional state (as in Fig. 2.14).

Recoverability RC(t) may be defined as the ease of restoring the service after failure. It can be modeled as:

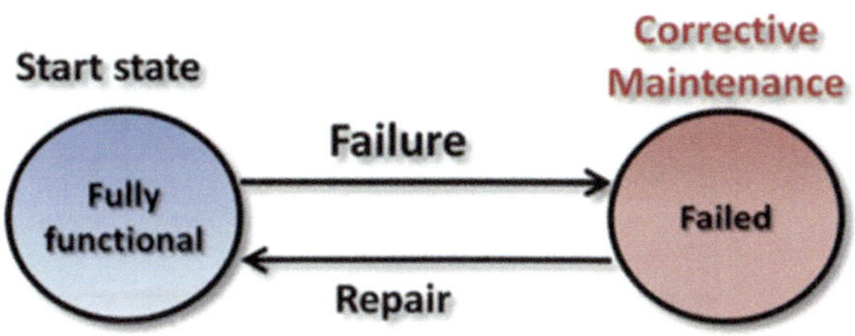

Fig. 2.14 A repairable system with two states and corrective maintenance

Equation 2.38 Recoverability of a system

$$\mathrm{RC}(t) = 1 - \mathrm{e}^{-\mu t} \tag{2.38}$$

where μ is the repair rate or average number of repairs that can be performed per time unit, the key aspects of recoverability, MTTR and MTTD, are given by:

Equation 2.39 Mean time to repair (MTTR) and mean time to detection (MTTD) of a system

$$\mathrm{MTTD} + \mathrm{MTTR} = \frac{1}{\mu} \tag{2.39}$$

Note that good testability would affect recoverability to a degree. MTTR will be further discussed below in the availability section.

2.5.2.2 Serviceability and Testability, $T(t)$

Testability $T(t)$ is the ease in which servicing and inspections can be conducted in order to identify the characteristics of a system; it is the ability to check certain attributes within a system. Measures of testability allow the system to assess the ease of performing tests. Ideally, in order to improve testability the tests can be automated and implemented as an integral part of the system.

These techniques can be used for error detection and error correction within the system. Since most of the time, testability is often used to determine the source of the problem, one way to improve the maintainability of the system significantly is the use of automatic diagnosis.

Testability relates to reliability since it allows detection and correction of errors that would, otherwise become failures, thus improving the overall reliability of the system. Testability is clearly connected with recoverability due to the importance of minimizing the time to locate and identify specific problems.

Two properties/measures closely associated to testability, *controllability* and *observability* (Franklin and Saluja 1995, p. 199; Goldstein 1979). *Observability* relates to the probability of "*observing*", via output measurements, the state of a system.

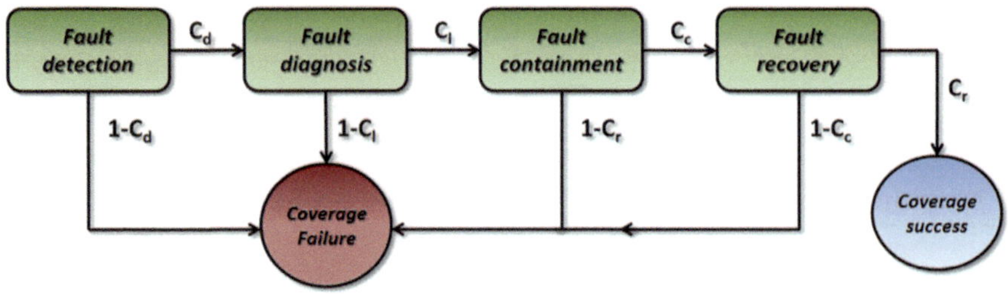

Fig. 2.15 Four phases of fault handling and their coverage

Controllability instead is associated to the ease of forcing parts of the system into desired states by using appropriate control signals. Design for testability techniques (DFT) (Alanen and Ungar 2011; Karimi and Lombardi 2002; Landis 1989; Mathew and Saab 1993), can be used in order to increase observability and controllability of systems.

2.5.2.3 Coverage

Mathematically, *fault coverage C* is the conditional probability that, given de existence of a fault in the operational system, the system is able to recover, and continue information processing with no permanent loss of essential information (Bouricius et al. 1969) i.e.:

Equation 2.40 Mathematical definition of coverage

$$C = \Pr[\text{system recovers} \mid \text{system fails}] \tag{2.40}$$

Fault coverage is a good measure of maintainability and, specifically of the system's ability to detect, locate, diagnose, contain and recover from the presence of a fault. Several types of fault coverage can be distinguished, depending on whether the designer is concerned with fault detection, diagnosis, containment or recovery (Kaufman and Johnson 2001).

In Fig. 2.15, we extend the phases of fault handling by Dugan and Trivedi (1989), showing the relationship among the steps of recovery and their coverage.

Fault detection coverage C_{d} measures the system's ability to detect fault. *Fault diagnosis coverage*[4] C_{l} is a measure of the system's ability to locate and determine the type of fault. *Fault containment/isolation coverage* C_{c} is a measure of the system's ability to contain faults within a predefined boundary (*fault containment region* or *FCR*).

[4] Fault diagnosis involves both the location (*fault location*) and determination of the fault type (*fault determination*).

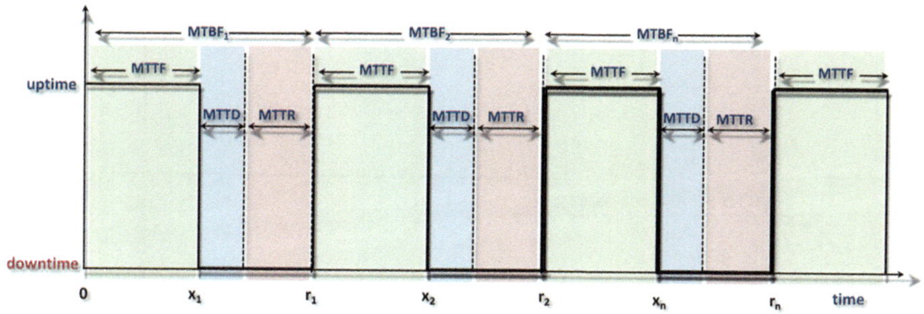

Fig. 2.16 Failure and repair cycle of a system

For instance, fault that occurs in a subsystem can be detected, located, and its effects can be prevented from propagating to other subsystems.

Finally, the general term "*coverage*" or "*fault coverage*" is often used to refer to *fault recovery coverage*, which measures the system's ability to recover from faults and maintain correct operation.

Recovery may involve modifying the structure to remove the faulty component (reconfiguration) including graceful degradation. The fault coverage C for the system is given by:

Equation 2.41 Coverage as a function of fault detection, diagnosis, containment and recovery coverages

$$C = C_{\mathrm{d}} \times C_{\mathrm{l}} \times C_{\mathrm{c}} \times C_{\mathrm{r}} \tag{2.41}$$

Clearly, high fault recovery coverage requires high fault detection, diagnosis and containment coverage.

2.5.3 Availability

A simple definition for *availability* of a repairable[5] system is "*Readiness for correct service*" (Avizienis et al. 2004). This measure is suitable for applications in which continuous performance is not essential but where it would be costly to have long downtimes. Availability is strongly dependent on how frequently the system becomes non-operational (reliability) and how quickly it can be repaired (maintainability) (see Fig. 2.16).

As defined in the MTBF equation (Eq. 2.22) the mean time between failures of a system can be defined as a combination of MTTF, MTTR and MTTD. Figure 2.16 illustrates the variations of the state (functional-failed) of a repairable system.

[5] The concept of availability is applicable to repairable systems. Availability of a non-repairable system would be the equivalent to reliability.

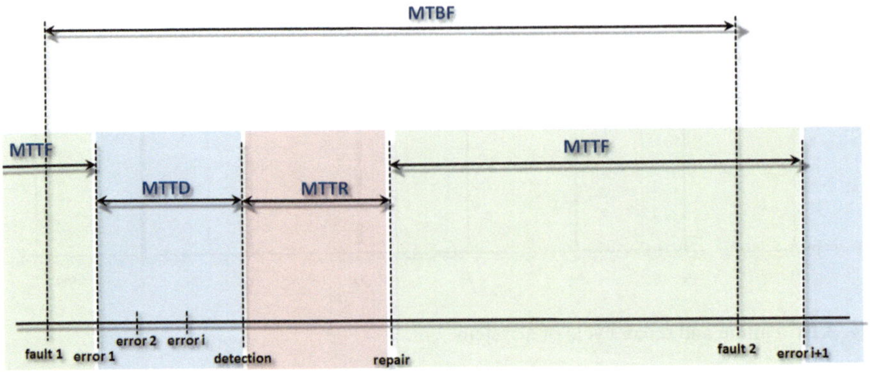

Fig. 2.17 Relation between time to failure (TTF), time between failures (TBF) and time to repair (TTR)

The time of operation of such systems is discontinuous. From time 0 to time x_1 the system is continuously available and therefore has an internal availability of 1.

After the first failure at time x_1 internal availability keeps decreasing until the detection and recovery mechanisms complete the repair at time r_1, returning to the original functional state. The system will fail again at time x_2 after a certain time of operation $[r_1 - x_2]$, get repaired at time r_2, and this process will reiterate. Assuming that Xi is an average of system failure and i an average of system repair, for $i > 1$:

Equation 2.42 Mean time between failures with average failure and system repair

$$\text{MTBF} = \sum_{k=1}^{n} (X_i - X(i-1)) \tag{2.42}$$

The relation between time to failure, time between failures and time to repair is displayed in Fig. 2.17.

There are various availability measures that can be classified differently depending on the time interval preferred or the downtimes used.

2.5.3.1 Instantaneous or Point Availability, $A(t)$

Instantaneous or *point availability* $A(t)$ is the probability that the system will be operational at a random time t. It describes the on-demand probability of proper service. It is equivalent to reliability when there is no repair.

While internal availability is based on an interval time, instantaneous availability is based on a specific instant of time. At any given time t, the system will be functional if one of the following conditions is met (Elsayed 1996):

1. The system was functional from 0 to t (it never failed by time t). The probability of this happening is $R(t)$ (Eq. 2.13).

2. The system has been functional since the last repair time r_i (see Fig. 2.16) when $0 < r_i < t$. This has a probability of:

Equation 2.43 Probability that the system has been functional since last repair time for $0 < r_i < t$

$$\int_0^t R(t - r_i)m(r_i)\mathrm{d}r_i \tag{2.43}$$

- With $m(r_i)$ being the renewal density function of the system.

The instantaneous availability of the system is the summation of these two probabilities:

Equation 2.44 Instantaneous or point availability of a repairable system

$$A(t) = R(t) + \int_0^t R(t - r_i)m(r_i)\mathrm{d}r_i \tag{2.44}$$

2.5.3.2 Average Uptime Availability (or Mean Availability), $\overline{A(t)}$

The average uptime availability or mean availability $\overline{A(t)}$ (Lie et al. 1977) is the proportion of time during a time period [0 − t] that the system is functional, this is given by:

Equation 2.45 Average uptime availability of a repairable system

$$\overline{A(t)} = \frac{1}{t}\int_0^t A(r_i)\mathrm{d}r_i \tag{2.45}$$

This type of measure is suitable to systems with periodical downtime for maintenance/repairing.

2.5.3.3 Limiting or Steady-State Availability, $A(\infty)$

The *limiting* or *steady state* availability (Applebaum 1965) of the system $A(\infty)$ is the limit of the availability function as time t tends to infinity:

Equation 2.46 Average up-time availability of a repairable system

$$A(\infty) = \lim_{t \to \infty} A(t) \tag{2.46}$$

2.5.3.4 Inherent Availability, A_I

In its simplest form, availability A can be mathematically generalized as a function of uptime and downtime

Equation 2.47 General availability as a function of uptime and downtime

$$A = \frac{\text{Uptime}}{\text{Uptime} + \text{Downtime}} \tag{2.47}$$

During the design phase of a FT system, *Inherent availability* A_I is a useful measure (Valstar 1965). A_I defines the availability of a system only in regard to effective functional time (uptime) and downtime due to *corrective maintenance (CM)*. It can be calculated using estimated parameters (MTTF, MTTD and MTTR) as:

Equation 2.48 Inherent availability as MTTF and MTBF

$$A_I = \frac{\text{MTTF}}{\text{MTTF} + \text{MTTD} + \text{MTTR}} = \frac{\text{MTTF}}{\text{MTBF}} \tag{2.48}$$

Hence, if MTTF or MTBF are long compared to MTTR and MTTD then the system's availability will be high. Likewise, if MTTR and MTTD are short then the system's availability will also be high. As reliability decreases (e.g. low MTTF), better recoverability will be needed (lower MTTR/MTTD) to achieve the same availability.

2.5.3.5 Achieved Availability, A_A

A_I is a good parameter to measure systems under ideal conditions where downtime due to *preventive maintenance* (PM) is overlooked. Achieved availability A_A is similar to inherent availability with the exception that downtimes due to PM tasks are also included. In can be defined as:

Equation 2.49 Achieved availability according to USA department of defence

$$A_A = \frac{\text{OT}}{\text{OT} + \text{TCM} + \text{TPM}} \tag{2.49}$$

where OT is the total operating time, TCM is the total corrective maintenance time and TPM the total time spent during preventive maintenance actions.

Table 2.1 Reliability-recoverability-availability relationship

Reliability	Recoverability	Availability
Constant	Constant	Constant
Constant	Decreases	Decreases
Constant	Increases	Increases
Decreases	Constant	Decreases
Increases	Constant	Increases

2.5.3.6 Availability-Recoverability-Reliability Relationship

At first glance, it might seem that a highly available system would also have high reliability. Nonetheless, this in not always the case, a system can be highly available yet suffer from frequent periods of non-operation as long as the length of the downtime is extremely short. Let's explore further the relationship between availability and reliability.

Reliability represents the probability of systems and components to perform its intended function for a desired period of time [0, t] under specified environmental and operating conditions. However, reliability in itself does not take into account any repair actions. Reliability does not reflect how long the recovery of a component/system will need in order to take it back to a working condition.

Availability reflects not only how often a system fails but how often it can be repaired (it accounts for repair actions). Thus, it is a function of reliability, recoverability and testability.

Equation 2.50 Availability as reliability and recoverability

$$A(t) = f(R(t), M(t)) \tag{2.50}$$

Table 2.1 presents the relationship between reliability, recoverability and availability. As shown by the table, once again, high reliability does not necessary imply high availability. Availability decreases as time to repair increases. Even an unreliable system could present high availability if MTTR is low.

2.6 Performability

The all-or-none nature of operation implicit in classic reliability and availability models does not measure in detail systems that can operate with different capability levels (e.g. multiprocessor systems).

Consequently, another key attribute of resilience, performability and its measure, *mean computation before failure (MCBF)* can be employed. MCBF is described as the expected amount of computation available on the system before its first failure, given an initial state (Beaudry 1978).

In qualitative terms, performability is the ability of a system or component to accomplish its designated functions within specified constraints such as speed, accuracy or memory usage. It is the measure of the likelihood that some subset of the functions of the system or component is performed correctly during a certain time interval.

Quantitatively, Performability $P(L, t)$ has been defined as "*the probability that the system's performance will be at or above some level L at the instant of time t*" (Fortes and Raghavendra 1985).

After the occurrence of faults and errors, certain systems have the ability to continue to perform correctly, however with a diminished level of performance.

This ability or feature is called *Graceful degradation*, or *fail soft operation* (Gountanis and Viss 1966), and it is the ability of a system (*gracefully degrading system* or *GDS*), upon failure of one or more of its component units, to continue the processing of tasks at the expense of decreasing its performance level.

The performability of a GDS $P(L, t)$ at time t depends on the amount of available resources and their computational capability provided.

Note that performability differs from reliability in that reliability measures the likelihood that all functions are performed properly, whereas performability measures the likelihood that some subset of the functions is performed properly.

2.7 Resilience

Historically, the term resilience has had multiple meanings in various fields. As a property it has different connotations. In social psychology resilience is about elasticity, spirit, resource and good mood. On the other hand, in material science resilience involves not only elasticity but robustness.

In computer science it has been identified as a synonym for fault tolerance. In this thesis we will adopt to extend the concept of resilience for safety critical applications. First we start by selecting the material science connotations. Hence, our definition of resilience includes both attributes: *robustness* and *elasticity*.

Figure 2.18 illustrates the different attributes and measures of resilience. The term robustness involves the use of static techniques such the use of very reliable materials or the use of rigid and pre-design approaches of fault tolerance. A robust system can deliver correct service in conditions beyond the normal domain of operation without fundamental changes to the original system.

This is more an aim that an objective. Total reliability to unforeseen faults other than the normal domain of operation is not feasible.

On the other hand, we interpret elasticity as the ability to spring back without losing the intrinsic properties of the material. Applied to resilience, we understand elasticity as the ability to *evolve*, to successfully accommodate changes (*evolvability*). An *evolvable system* may perform changes to the system, decreasing its level of performance or reliability for a specific time range (1) to compensate for faults or (2) during exceptional circumstances (graceful degradation).

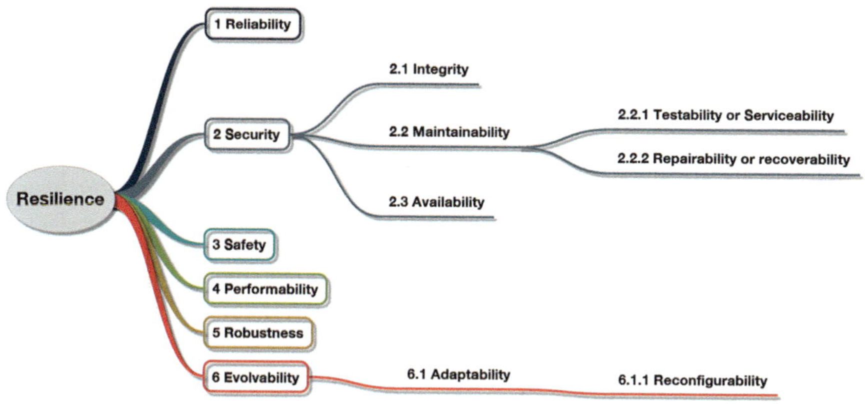

Fig. 2.18 Attributes and measures of resilience

More specifically, we consider that a resilient system must have the ability to be adaptable, understanding *adaptability* as the ability to evolve while executing. Therefore, adaptability is a subset of evolvability and requires the ability to anticipate to changes prior to the occurrence of the resulting damage.

Therefore a resilient architecture must include different mechanisms to acquire both attributes: (a) static pre-design fault tolerant techniques (robust) and (b) dynamic techniques (elastic) that may be achieved with the ability to reconfigure elements of the system (reconfiguration).

2.7.1 Requirements

The main aim required to implement a resilient architecture for safety critical applications is the "ability to deliver correct service adapting to disturbance, disruption and change within specified time constraints".

The above aim can be subdivided into more specific objectives, as follows:

1. Continuity of service (reliability);
2. Readiness for usage (availability);
3. Non-occurrence of catastrophic consequences (safety);
4. Non-occurrence of incorrect system alterations (integrity);
5. Ability to undergo corrective maintenance and recovery with maximum coverage of faults (testability and recoverability);
6. Ability to perform in the presence of faults (performability);
7. Ability to decrease the level of performance for a specific time range in order to compensate for hardware faults (graceful degradation);
8. Ability to regain operational status via reconfiguration in the presence of faults (recoverability via reconfiguration);

9. Ability to accommodate changes (evolvability);
10. Ability to anticipate to changes (adaptability).

2.7.2 *Effectiveness of Resilience*

Being reliability $R(t)$, security $\mathrm{SC}(t)$, integrity $I(t)$, maintainability $M(t)$, testability $T(t)$, recoverability $\mathrm{RC}(t)$, availability $A(t)$, safety $S(t)$, performability $P(L, t)$, robustness $\mathrm{RB}(t)$, evolvability $E(t)$, adaptability $\mathrm{AD}(t)$ and reconfigurability RC(). Maintainability is a function of serviceability and repairability:

Equation 2.51 Maintainability as a function of serviceability and repairability

$$M(t) = f(T(t), \mathrm{RP}(t)) \tag{2.51}$$

Security is a function of integrity, availability and maintainability:

Equation 2.52 Security as a function of Integrity, availability and maintainability

$$\mathrm{SC}(t) = f(I(t), A(t),\ M(T(t),\ \mathrm{RP}(t))) \tag{2.52}$$

Evolvability is a function of adaptability and reconfigurability:

Equation 2.53 Evolvability as a function of adaptability and reconfigurability

$$E(t) = f(\mathrm{AD}(t), \mathrm{RC}(t)) \tag{2.53}$$

Therefore, resilience $\mathrm{RES}(t)$ will be a function of all these attributes:

Equation 2.54 Availability as reliability and recoverability

$$\mathrm{RES}(t) = f\begin{pmatrix} \text{reliability, integrity, testability, recoverability,} \\ \text{availability, safety, performability,} \\ \text{robustness, adaptability, reconfiguration} \end{pmatrix} \tag{2.54}$$

With all these attributes the following systems would benefit from the implementation of effective resilience:

1. Safety-life critical: e.g. aircraft and nuclear reactor control, life support systems;
2. Business critical;
3. Reliable critical: e.g. telephone switching-, traffic light control-, automotive control (ABS, fuel injection) systems;
4. Mission critical and long life systems: e.g. manned and unmanned space borne, satellites and other systems in inaccessible locations;
5. Transaction processing and non-stop systems that demand high availability

Resilience is not a simple and single concept, rather, it possesses different components or key attributes. Taking into consideration all these attributes, our definition of resilience is as follows:

> A resilient system is a system that over a specified time interval, under specified environmental and operating conditions, is ready to perform its intended function, guaranteeing the absence of improper system alterations, having the ability to anticipate and accommodate changes while executing, and the ability to conduct servicing and inspections so that in case of failure quick restoration to a specified working condition must be achieved, or otherwise discontinue of the operation in a safe way is provided.

2.8 Conclusion

The chapter explains the concept of resilience that encompasses important attributes and measures that will be used through the whole book.

Safety critical systems must provide correct service at all times by avoiding the occurrence of any catastrophic failure. Different techniques can be employed to increase reliability by avoiding/preventing hardware faults from becoming errors that may lead to failures and catastrophic failures.

We present the concept of *vicious cycle* that explains the reasons behind the performance and reliability problems that the microprocessor industry is currently facing. The increase of transistor density, operating frequencies and architectural complexity is drastically decreasing the reliability of newer systems. There is, therefore, a need for implementing mechanisms that can deal with the upcoming fault rates.

The basic definitions for reliability evaluation have been reviewed. It has been shown that for constant failure rate, independent of time, the exponential distribution is the most suitable for the reliability analysis of the useful time of systems. The age of a system should be taken into account when analyzing reliability. Three different periods with different reliability distributions have been explained by introducing the bathtub curve, which represents very well the effect that ageing and degradation have on HW reliability.

Introducing deliberately and carefully extra components into a system, overall reliability can be increased, provided reliability benefit accruing from the redundancy scheme exceeds the decrease in reliability due to the actual implementation of the redundancy mechanisms itself.

We extend the classical definition of resilience and apply it to the field of safety critical computing. Moreover, we quantify the key attributes that a resilient system must have, exploring the relationships among these quantitative measures. The attributes of safety and performability are explained. The concept of security is described, including its attributes: integrity, availability, testability and recoverability.

We show how the availability of FT systems can be estimated using different methods and measures.

A resilient system, over a specified time interval, under specified environmental and operating conditions (performability), "must be ready" (in terms of availability) to perform its intended function (reliability), guaranteeing the absence of improper system alterations (integrity).

It must have the ability to conduct servicing and inspections (testability) so that in case of failure quick restoration to a specified working condition must be achieved (maintainability) can be provided or can discontinue its operation in a safe way (safety).

Furthermore, a resilient system must have the ability to anticipate changes and evolve (evolvability) while executing (adaptability), successfully accommodating changes by reconfiguring elements of the system.

Acknowledgements Our endless debates on the impacts of radiation on electronics for safety-critical applications with Dr Victor Castano made this chapter possible.

References

Alanen J, Ungar LY (2011) Comparing software design for testability to hardware DFT and BIST. In: 2011 IEEE AUTOTESTCON, pp 272–278

Applebaum SP (1965) Steady-state reliability of systems of mutually independent subsystems. IEEE Trans Reliab 14:23–29

Asanovic K, Bodik R, Catanzaro B, Gebis J, Husbands P, Keutzer K, Patterson D, Plishker W, Shalf J, Williams S, Yelick K (2006) The landscape of parallel computing research: a view from Berkeley. Technical report No. UCB/EECS-2006-183

Avizienis A (1982) The four-universe information system model for the study of fault tolerance. In: Proceedings of the 12th annual international symposium on fault-tolerant computing, Santa Monica, California, pp 6–13

Avizienis A, Laprie J-C, Randell B, Landwehr C (2004) Basic concepts and taxonomy of dependable and secure computing. IEEE Trans Depend Secure Comput 1:11–33

Beaudry MD (1978) Performance-related reliability measures for computing systems. IEEE Trans Comput 27:540–547

Birolini A (2014) Reliability engineering, 7th edn. Springer. ISBN: 978-3-642-39534-5

Bodsberg L, Hokstad P (1995) A system approach to reliability and life-cycle cost of process safety-systems. IEEE Trans Reliab 44:179–186

Bouricius WG, Carter WC, Schneider PR (1969) Reliability modeling techniques for self-repairing computer systems. In: Proceedings of the 1969 24th national conference, ACM'69. ACM, New York, NY, USA, pp 295–309

Dhillon BS (2006) Maintainability, maintenance, and reliability for engineers. CRC Press

Dreslinski RG, Wieckowski M, Blaauw D, Sylvester D, Mudge T (2010) Near-threshold computing: reclaiming Moore's law through energy efficient integrated circuits. Proc IEEE 98:253–266

Dugan JB, Trivedi KS (1989) Coverage modeling for dependability analysis of fault-tolerant systems. IEEE Trans Comput 38:775–787

Elsayed EA (1996) Reliability engineering. Prentice Hall

Fortes JAB, Raghavendra CS (1985) Gracefully degradable processor arrays. IEEE Trans Comput 34:1033–1044

Franklin M, Saluja KK (1995) Embedded RAM testing. In: Records of the 1995 IEEE international workshop on memory technology, design and testing, pp 29–33

Gnedenko B, Pavlov IV, Ushakov IA (1999) Statistical reliability engineering, 1st edn. Wiley-Interscience

Goldstein L (1979) Controllability/observability analysis of digital circuits. IEEE Trans Circuits Syst 26:685–693
Gountanis RJ, Viss NL (1966) A method of processor selection for interrupt handling in a multiprocessor system. Proc IEEE 54:1812–1819
Hutcheson GD (2009) The economic implications of Moore's law. In: Into the nano era, pp 11–38
IBM (2007) IBM Power6 microprocessor and IBM system, 21 May 2007, p 570
Johnson BW (1989) The design and analysis of fault tolerant digital systems. Addison-Wesley
Karimi F, Lombardi F (2002) A scan-BIST environment for testing embedded memories. In: Proceedings of the eighth IEEE international on-line testing workshop, 2002, pp 211–217
Kaufman L, Johnson BW (2001) Embedded digital system reliability and safety analysis. NUREG/GR-0020
Koren I, Krishna CM (2007) Fault-tolerant systems. Morgan Kaufmann
Kovalenko IN, Kuznetzov NY, Pegg PA (1997) Mathematical theory of reliability of time dependent systems with practical applications, 1st edn. Wiley
Landis DL (1989) A self-test system architecture for reconfigurable WSI. In: Proceedings. Meeting the tests of time, international test conference, 1989, pp 275–282
Laprie JC (2008) From dependability to resilience. In: 38th IEEE/IFIP international conference on dependable systems and networks, June 2008, pp G8–G9
Laprie J, Avizienis A (1986) Dependable computing: from concepts to design diversity. In: Proceedings of the IEEE, pp 629–638
Laprie JCC, Avizienis A, Kopetz H (eds) (1992) Dependability: basic concepts and terminology. Springer-Verlag, New York
Laprie J, Kopetz H, Littlewood B (eds) (1995) Predictably dependable computing systems. Springer-Verlag Heidelberg, pp 1–24. ISBN-13: 978-3-64279791-0
Lie CH, Hwang CL, Tillman FA (1977) Availability of maintained systems: a state-of-the-art survey. AIIE Trans 9:247–259
Mathew B, Saab DG (1993) Partial reset: an inexpensive design for testability approach. In: 4th European conference on design automation, 1993 proceedings, pp 151–155
Nair R (2002) Effect of increasing chip density on the evolution of computer architectures. IBM J Res Dev 46:223–234
Patterson D, Hennesy J (2012) Computer architecture: a quantitative approach. Elsevier. ISBN: 978-0-12-383872-8
Pierce WH (1965) Failure-tolerant computer design. Academic Press, New York
Schagaev I, Gutknecht J (2024) Software design for resilient computer systems. https://doi.org/10.1007/978-3-031-55139-0. ISBN: 978-3-031-55138-3
Shivakumar P, Kistler M, Keckler SW, Burger D, Alvisi L (2002) Modeling the effect of technology trends on the soft error rate of combinational logic. In: Proceedings of the international conference on dependable systems and networks, 2002 (DSN 2002), pp 389–398
Valstar JE (1965) The contribution of testability to the cost-effectiveness of a weapon system. IEEE Trans Aerosp 3:52–59
Von Neumann J (1956) Probabilistic logics and the synthesis of reliable organisms from unreliable components. In: Automata studies. Princeton University Press, pp 43–98

Chapter 3
Dealing with Faults: Redundancy

Abstract This chapter provides a survey of state-of-the art design strategies to handle faults with special stress on redundancy-based techniques. Section one presents an overview of fault avoidance design strategies. Section two provides a survey of fault tolerance techniques. The notion of redundancy and its different types are presented and a notation, which may be used to describe the different types of redundancy, is introduced. The concepts, capabilities and applications of the different techniques based on structural (SR), information (IR) and time redundancy (TR) are compared and discussed.

Keywords Redundancy · Reliability · Structural redundancy · Informational redundancy · Time redundancy · Error detection

3.1 Handling Faults: Design Strategies

In order to increase the reliability of safety-critical systems so that correct service can be delivered, techniques need to be developed to prevent or reduce the appearance of faults that could cause catastrophic failures. Depending on the phase of the development cycle and the level of abstraction at which the faults are tackled, two different design strategies can be adopted: fault avoidance and fault tolerance.

Fault avoidance strategies can be used at device level during design time. Typical in mainstream applications, in order to reduce the number of failures, this approach focuses on preventing the occurrence of faults.

Since a failure is the consequence of an error propagating, and an error is the consequence of a fault, eliminating faults would maximize reliability. Examples of this are silicon on insulator (SOI) and hardened memory cells. These techniques have drawbacks in terms of cost, speed of operation and chip area.

At execution or run time and at different levels of abstraction, fault-tolerance strategies can be implemented.

© The Author(s), under exclusive license to Springer Nature Switzerland AG 2026

I. Schagaev, *Resilience of Computer Systems*,
https://doi.org/10.1007/978-3-032-00938-8_3

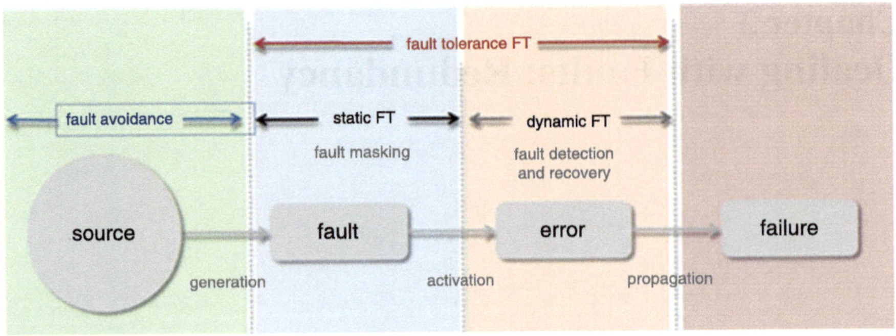

Fig. 3.1 Mechanisms to deal with faults within the fault-failure lifecycle

Following the failure lifecycle and its different phases already described in Sect. 2.1, Fig. 3.1 adds the different mechanisms to deal with faults within the fault generation, error activation and failure propagation phases.

Additionally, Fig. 3.1 serves as a summary of the chapter introducing the fault avoidance and fault tolerance techniques and their phase of interaction within the failure lifecycle.

Focusing on the source of faults, fault avoidance mechanisms attempt to prevent faults from occurring in the first place. Once a fault has been generated it can be prevented from activating an error using static fault tolerant techniques such as masking.

Alternately, errors can be detected and recovered using dynamic fault tolerance techniques. Therefore, either we prevent the faults from taking place (fault avoidance) or we deal with them using fault tolerance techniques.

3.2 Fault Avoidance

Nowadays, mainstream systems employ fault avoidance design strategies in order to achieve their projected failure rates. Manufacturing companies perform assessments of sources and weaknesses that could lead to potential failures. Based on the assessments, preventive measures are taken to ensure that the overall reliability target is not compromised.

Additionally, fault avoidance strategies may include technology and design mitigation techniques that implicate modifications of conventional manufacturing processes. These techniques involve the use of specific materials, the modification of the doping profiles of devices and substrates and the optimization of deposition processes for insulators.

Technology mitigation techniques consist of IC process variations by either improving the manufacturing process or by improving the materials used.

Improving materials: implicates the selection of specific materials with better characteristics, e.g.:

1. Boron has been used extensively as a p-type dopant in silicon and has also been used in Boron-Phosphor-Silicate-Glass (BPSG) dielectric layers. For BPSG-based semiconductor processes, BPSG can, in fact, be the predominant source of transient errors (Baumann 2001). The removal of B-10 Boron isotopes in BPSG has been proven effective in the reduction of transient errors (Baumann et al. 1995).
2. *Lead-free* materials can reduce the effect of alpha particles (May 1979) (extensive information on alpha particle effects is provided in Chap. 4).
3. Implanting of elements such as Al, As, Fl, P, and Si into oxides improves the resilience to Total Ionizing Dose effects (TID) (Kato et al. 1989; Mrstik et al. 2000; Nishioka et al. 1989).

Improving the manufacturing process is based on changing the charge collection and charge sharing capabilities of the devices:

1. *Substrate + techniques*: e.g. using epitaxial substrate doping (EPI layer charge reduction) (Puchner et al. 2006), wells (single well, twin well and triple well processes) (Pellish et al. 2006; Puchner et al. 2006; Roche and Gasiot 2005), buried layers (Roche and Gasiot 2005) and dry thermal oxidation (Hughes and Benedetto 2003).
2. *Non-capacitance techniques*: e.g. increasing the node coupling capacitance between storage nodes and memory, or using a DRAM capacitor on top of the memory cell.
3. Using *alternative insulating substrates*; e.g. the use of Silicon on Insulator (SOI) or Silicon on Sapphire (SOS) (Schwank et al. 2003) would mitigate significantly the transient faults due to radiation (described in Chap. 4).

Whilst technology mitigation techniques are based at the process level, **design mitigation techniques** operate at the layout level. An example of this type of technique is the use of enclosed layout transistors. Furthermore, to prevent the effects of radiation memory cells can be hardened with the use of contact and guard rings.

The effect of silicon failure mechanisms, such as radiation induced transient faults and wear-out defects, is proportional to the clock speed, supply voltage, temperature, etc. Therefore, to ensure system reliability safety margins are inserted into clock speed, operating temperature and supply voltage margins. If the system failure rates resulting from the use of fault avoidance strategies fall within the specified reliability targets, the use of redundancy techniques is not justified. However, this is not the case for safety-critical systems.

Despite all the testing, verification techniques and technology improvement, hardware components will eventually fail. The fault avoidance approach will not be panacea and will be insufficient if:

1. Failure rate and MTTR are unacceptable or,
2. The system is inaccessible for repair and maintenance actions.

Therefore, fault avoidance techniques are only part of the solution for real time safety critical domains. Complete removal of faults via fault avoidance is not possible; above all, it has drawbacks in terms of cost of manufacturing the elements required, speed of operation and increased chip area.

3.3 Fault Tolerance: Using Redundancy

The key ingredient of fault tolerance is redundancy. Redundancy is defined as the addition of information, resources or time beyond what is needed for correct system operation (Schagaev and Sogomonian 1988; Latchoumy et al. 2011).

Fault tolerant techniques rely on redundancy that may include a combination of additional elements of hardware and/or software to detect and/or recover from faults. These components are called redundant since they are not required in a *perfect + system.*[1]

Artificially built-in or *protective redundancy* is a system property that we define as the incorporation of extra components (transistors at a low level) in the design of a system so that its function is not impaired in the event of a failure. Redundancy may arise by design (artificially built-in redundancy) or as a natural by-product of design (*natural redundancy*).

Natural redundancy is usually unexploited whilst artificially built-in redundancy has been deliberately introduced. In this thesis, when the term redundancy (or redundant) is used it is meant to have the artificial connotation instead of the natural one.

When a system does not provide the minimum reliability required, extra redundancy, not strictly necessary for the normal functioning of the system can be added in order to increase the probability of normal functioning.

Notice that that the term redundant does not mean identical functionality; it just denotes that it performs the same task. In this sense, heterogeneous hardware performing the same work can also provide redundancy.

Fault tolerance assumes actions such as fault detection, location of the faulty component, recovery, and if necessary, reconfiguration of the system. Fault detection is the process of determining the presence of faults and the time of occurrence.

Fault location is to exactly locate the reason/origin of the fault. The system must be dynamically restored as though it is "*as good as new*" in operational terms, except for the fact that some of the redundancy has been used up and this may limit the possibilities for future repairs.

Many different attempts to classify redundancy have been made (Carter and Bouricius 1971; Schagaev 1989, 2001). This work follows the classification of redundancy of fault-tolerant systems approach proposed by Schagaev (1989, 2001).

[1] A perfect system is a system with a theoretical 100% reliability. A perfect system is usually assumed to model extra reliable systems.

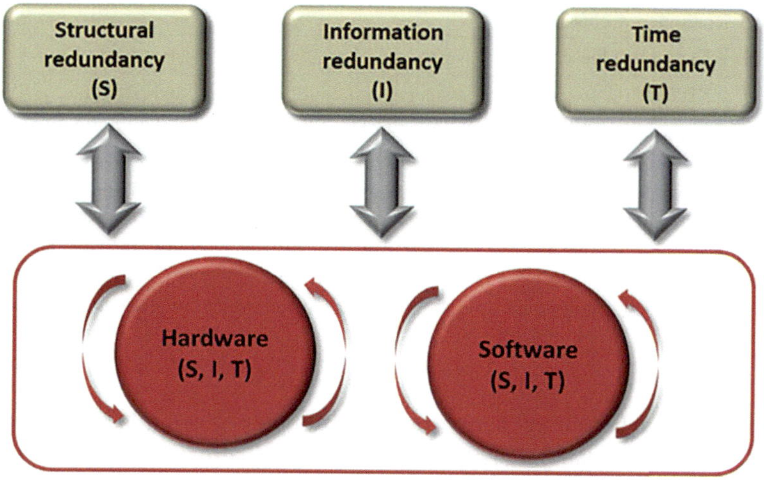

Fig. 3.2 Redundancy types and their implementation

Figure 3.2 shows the different types of redundancy (at the top of the figure) and the ways it can be implemented using hardware and software (at the bottom of the figure). In general, three types of redundancy exist: structural (S), involving multiplication of components, information (I), involving multiplication of information/data and time redundancy (T), involving multiplication of functions in time. These can be implemented in hardware and software. We mostly focus on the hardware aspect of redundancy and fault tolerance.

Redundancy comes with a cost. Information and structural redundancies require additional hardware components, extra power and perhaps extra area and shielding. Time redundancy requires faster processing to achieve the same performance, which in turn requires extra hardware and power. Software redundancy involves higher development and maintenance costs.

3.3.1 Redundancy Notation

Existing implementations of system redundancy use at least one of these three redundancy types, usually more than one and can be implemented in hardware HW(), software SW() or a combination of both (HW(), SW()).

As an example, hardware based information redundancy is abbreviated as HW(I). Additional quantifiers are used together with the redundancy type to further specify the used redundancy as shown in Table 3.1.

Note that the current notation does not include the implementation level. HW($2S$) only indicates duplication, but not whether the whole system is duplicated or it is just parts of that system, such as, for example, duplicated memory.

Table 3.1 Redundancy classifiers

Quantifier	Example	Description
	SW(I)	No quantifier means general, not further specified redundancy. SW(I) for instance just indicates general software information redundancy
δ	SW(δI)	Additional used software-based redundancy
Number	HW($2S$)	The number indicates duplication (2), triplication (3), etc. of a system if used as a prefix for the redundancy type. The original system and the copies are identical. n instead of a discreet number is used to mark repetition until success in case of time redundancy
Indices	HW(S_1, S_2)	Indices are used to mark a duplicated system implementation/hardware components

Tables 3.2 and 3.3 present some concrete examples of notation of hardware and software-based redundancy. Any type of redundancy (hardware and software) needs additional structural redundancy for its implementation.

For instance, instruction repetition HW(nT) needs additional hardware registers to store the internal state to be able to perform instruction rollback. We refer to this as *supportive redundancy* and we define it as the redundancy needed for the implementation of the main redundancy technique. For the sake of simplicity, we usually omit this supportive redundancy.

In cases where it is not clear whether an applied redundancy type is supportive or not, more than one redundancy type can be used. An example of this is the case

Table 3.2 Examples of notation of HW based redundancy

Redundancy type	Description
HW($2S$)	Structural (material) redundancy of hardware such as duplicated memory system
HW(S_1, S_2)	A duplicated FT computer system with principally non identical parts
HW(δI)	Redundant information bit, for example an additional parity bit per data word in HW memory for error detection
HW(nT)	Special HW to delay execution (like in a timing diagram) to avoid transient faults
HW(δT)	Special HW to delay execution to avoid transient faults

Table 3.3 Examples of notation of SW based redundancy

Redundancy type	Description
SW($2S$)	Structural (material) redundancy of hardware, such as duplicated memory system
SW(S_1, S_2)	A duplicated FT computer system with principally non-identical parts
SW(δI)	Redundant information bit, for example an additional parity bit per data word in HW memory for error detection

of software based and hardware checks that are performed during idle time of the system: SW(δS, δT).

3.4 Structural Redundancy HW(*S*)

Structural hardware redundancy involves the multiplication of independent hardware components and execution of the same computation over such components at the same time. Errors are exposed by checking/comparing the results of the independent executions.

In terms of granularity, redundancy in general, and not only hardware redundancy, can be applied at different abstraction levels. From bottom up we can distinguish between finer-grained: transistor level, gate or logic level, and between coarser-grained designs: circuit level, function level, system level, microcode level and chip level of abstraction.

Therefore, redundant components can be as simple as transistors or logic gates but also as complex as processors or even larger entities.

Figure 3.3 displays a taxonomy of the different hardware techniques based on structural redundancy. Two different architectures of redundancy can be distinguished: Parallel redundancy with redundant components running concurrently and Standby redundancy with a spare component being activated upon failure of an active component.

Furthermore, these extra resources can be used passively (*passive redundancy*), actively (*active redundancy*) or combined.

In systems with active redundancy, all redundant components are in operation, sharing the load with the normal components. This implies that both, regular and redundant components, age together. Passive components are not fully energised

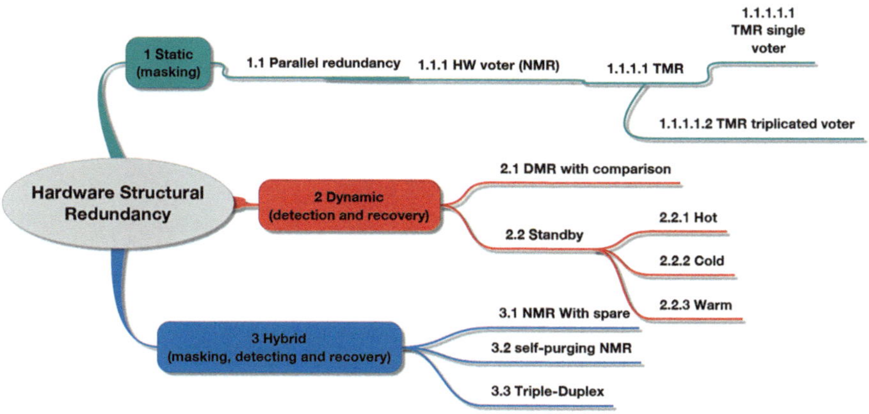

Fig. 3.3 Taxonomy of structural HW redundancy

and start normal operation only when normal components fail. Passive components can be further broken down into two types: *warm* and *cold standby*.

Warm standby components remain partially energised until becoming active and tend to deteriorate with time, hence, having lower failure rate than the regular components. *Cold standby* components are kept in reserve and they only become energized when put into use. These types of components have a zero failure rate, meaning they do not fail when they are in standby mode.

Whilst passive components are switched off completely, standby components are partially activated. Standby redundancy is usually applied when the start time of the component is unacceptably long.

3.4.1 Static Redundancy

Static redundancy, also called masking redundancy, implements error mitigation. The term static relates to the fact that redundancy is built into the system structure. Fault tolerant techniques based on this type of redundancy (*static fault tolerance*) transparently remove errors on detection.

The most common form of hardware redundancy is *Triple modular redundancy (TMR)* (Von Neumann 1956) and its generalization *N-modular redundancy (NMR)*. Note that *Dual modular redundancy (DMR)* (*DMR* is further explained in Sect. 3.4.2.1) is not considered static redundancy since the mismatch can take place but recovery is not possible.

3.4.1.1 Triple Modular Redundancy: HW($3S$) + HW(δS)

A basic TMR system (two-out-of-three) is a fault-tolerant form of NMR that consists of three fully redundant and active components or modules working in parallel with equivalent functionality (Johnson 1989; von Neumann 1956).

Figure 3.4 presents an example of a TMR system with a voter.

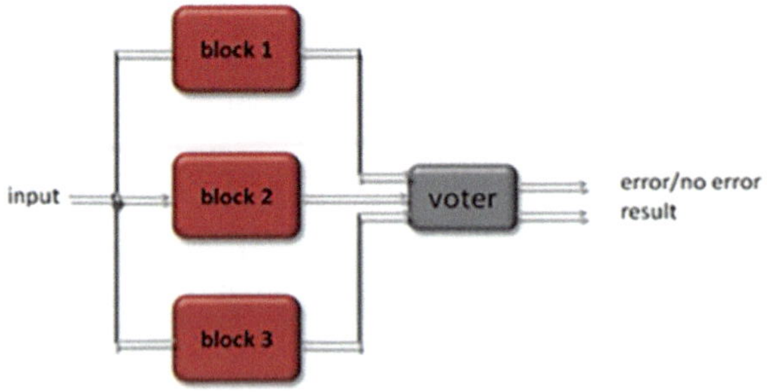

Fig. 3.4 Triple modular redundancy (TMR) with a voter

The three components perform a process based on individual inputs whose results are in turn processed by a voting system to produce a single output. The voting is based on majority; if any of the three components has a fault, the other two systems can mask the fault.

It is assumed that two out of three modules must deliver correct results. Therefore, *TMR* is capable of masking a single error.

Generally speaking, a majority voting mechanism should:

1. Guarantee a majority vote on the input data to the voter
2. Determine the faulty block.

In order to guarantee the majority vote, loosely synchronized systems require synchronization of the inputs to the voter.

A specific example of this technique is the *Boeing TMR 777* primary flight computer (Yeh 1996), which has triple redundancy for all hardware, including computing system, communication paths, electrical and hydraulic power.

3.4.1.2 Comparing the Reliability of Simplex and TMR with Perfect Voter Systems

A simple *TMR* system such the one in Fig. 3.4 includes three blocks, two of which are required for the system to provide correct service. Given the reliability of a single component or simplex system by:

Equation 3.1 Reliability of a simplex TMR system

$$R_{\text{simplex}} = \mathrm{e}^{-\lambda t} \tag{3.1}$$

where λ is the failure rate for the single component; the MTTF of a simplex system can be expressed as:

Equation 3.2 MTTF of a simplex TMR system

$$\text{MTTF}_{\text{simplex}} = \int \mathrm{e}^{-\lambda t} = \frac{1}{\lambda} \tag{3.2}$$

The reliability of a *TMR* system with a perfect voter is given by:

Equation 3.3 Reliability of a TMR system with a perfect voter

$$\begin{aligned} R_{\text{TMR}} &= R_{\text{m}}^3 + \binom{3}{2} R_{\text{m}}^2 (1 - R_{\text{m}}) \\ R_{\text{TMR}} &= \mathrm{e}^{-3\lambda t} + \binom{3}{2} \mathrm{e}^{-2\lambda t} \left(1 - \mathrm{e}^{-\lambda t}\right) \\ R_{\text{TMR}} &= 3\mathrm{e}^{-2\lambda t} \end{aligned} \tag{3.3}$$

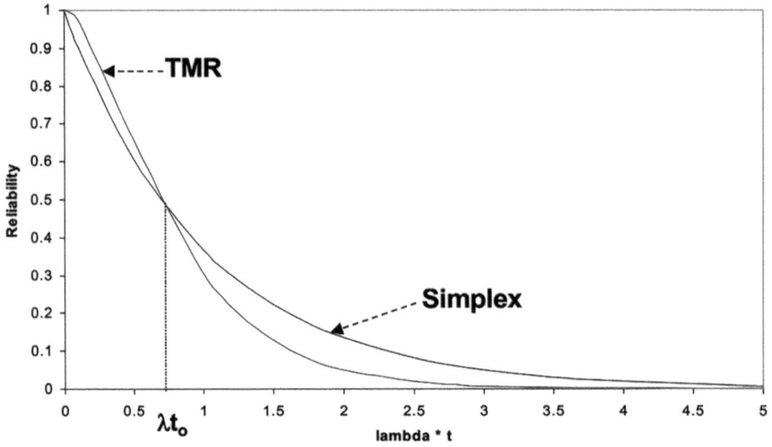

Fig. 3.5 TMR has higher reliability than simplex for short missions ($t < t_0$)

And, therefore:

Equation 3.4 MTTF of a TMR with a perfect voter

$$\begin{aligned} \mathrm{MTTF_{TMR}} &= \frac{3}{2\lambda} - \frac{2}{3\lambda} = \frac{5}{6\lambda} \\ \mathrm{MTTF_{simplex}} &> \mathrm{MTTF_{TMR}} \end{aligned} \tag{3.4}$$

Figure 3.5 presents comparison reliability for both: Simplex and TMR systems. One should note that:

Equation 3.5 Comparative reliability of TMR and simplex systems MTTF of a TMR with a perfect voter

$$\begin{aligned} R_{TMR}(t) &\geq R(t) \;\; 0 \leq t \leq t_0 \\ R_{TMR}(t) &\leq R(t) \;\; t_0 \leq t \leq \infty \end{aligned} \tag{3.5}$$

where:

$$t_0 = \frac{\ln 2}{\lambda} \approx \frac{0.7}{\lambda}$$

TMR is very useful in aircraft applications offering 0.99999 reliability over a 10-h period. Iyer (2003) shows that TMR is not suitable for long safety-critical missions ($t > t_0$) because paradoxically, after the first failure, the two remaining components compete to fail. Higher reliability can be achieved extending TMR to N-Modular Redundancy. Therefore, a blind use of redundancy can lead to seemingly paradoxical results.

3.4.1.3 Reliability of TMR with Voting

The previous expression of reliability of *TMR* assumes that the voter is perfect, that is to say that the voter is 100% reliable. The reliability of a generic TMR system with non-perfect single voting (*TMRV*) and identical blocks is given by:

Equation 3.6 Reliability of a *TMR* system with a non-perfect voter and identical blocks

$$R_{\mathrm{TMRV}} = R_{\mathrm{V}}\left(R_{\mathrm{m}}^{3} + \binom{3}{2}R_{\mathrm{m}}^{3}(1 - R_{\mathrm{m}})\right) \tag{3.6}$$

where R_{v} is the reliability of the voter mechanism and R_{m} is the reliability of the block. In terms of reliability, the voter becomes the weak part of this configuration.

The voter is a *single point of failure* (*SPF*); if the voter fails then the complete system will potentially fail. This can be tackled following different alternatives:

1. By increasing the reliability of the voter using fault avoidance techniques
2. By triplicating the voter and connecting the module outputs to all three voters (Johnson 1989), so that individual voting failures can be corrected by the extra voting process
3. By implementing online self-testing for the voting circuitry (Cazeaux et al. 2004; Metra et al. 1997)
4. Using an I_{DDQ} *checkable voters* (*ICVs*) (Bogliolo et al. 2000): under fault-free conditions, *ICVs* work as traditional *CMOS* voters; however, they cause *quiescent supply currents* (*IDDQs*)[2] in the presence of maskable stuck-at faults (see Sect. 5.3.1). Faults can be detected using I_{DDQ} *testing*, by monitoring *IDDQs* (Williams et al. 1996).

A basic TMR system does not support common-mode failures (*CMFs*[3]) (Lala and Harper 1994). *CMFs* are the result of failures affecting more than one component, usually due to a common cause, which may be due to design-faults or operational ones resulting from external (such radiation or electro-migration) or internal causes.

For instance, a radiation source causing multiple event upsets (Reed et al. 1997) can potentially lead to the failure of more than one component in a *TMR* system.

---

[2] Quiescent current is the current consumed by a circuit when no load is present. Fault-free CMOS devices have very quiescent currents when they are in a quiescent state. Faults that cause high quiescent currents can be detected if the quiescent current is significantly higher than that of a fault-free circuit (Williams et al. 1996).

[3] Multiple faults can be either independent (attributed to different causes) or related (attributed to a common cause). Both can lead to similar errors (e.g. errors that cannot be distinguished by the detection mechanisms being used) (Avizienis and Kelly 1984). The failures triggered by similar errors are called CMF.

3.4.1.4 N-modular Redundancy: HW(nS) + HW(δS)

The generalized version of *TMR* is *NMR*, where N stands for the number of redundant modules. The main advantage of using N modules as opposed to only three is that often more faults can be tolerated.

For instance, a *5MR system* contains 5 replicated modules including a majority voting arrangement. The voter allows the system to deliver correct service in case of as many as two module faults.

Figure 3.6 depicts a generic N-modular redundant system with a voter. The redundancy of this system can be defined as HW(nS) + HW(δS) using the previous notation. NMR works similarly to TMR but this type of structure is able to detect $[(N - 1)]/2$ errors in different processing modules.

Besides TMR, 5- and 7-modular redundancies are the most common structures and are capable of detecting two and three errors, respectively.

M-out-of-N systems are a type of NMR. The reliability of a generic M-out-of-N system, assuming that it has a perfect voter and M out of N modules need to function, is expressed by:

Equation 3.7 Reliability of an M-out-of-N System with perfect voter

$$R_{MN} = \sum_{i=0}^{N-M} \binom{N}{i} R_m^{N-i}(1 - R_m)^i \quad (3.7)$$

Note than *NMR* systems offer higher reliability than *TMR* but at a much higher cost. Undoubtedly, for practical applications, there must be some limit on the amount of redundancy that can be employed.

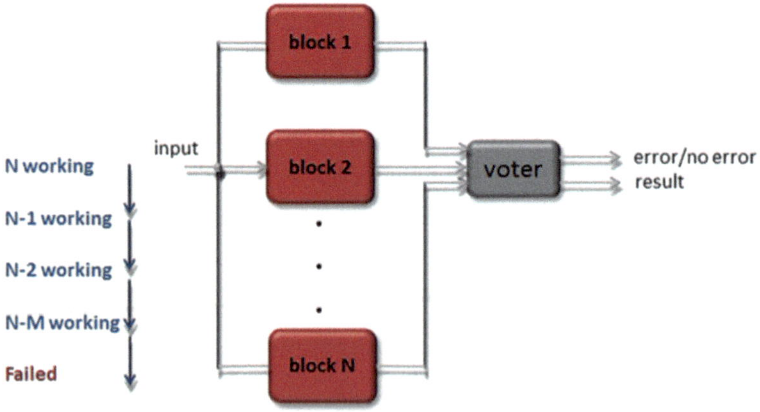

Fig. 3.6 N-modular redundancy with a voter: M-out-of-N system

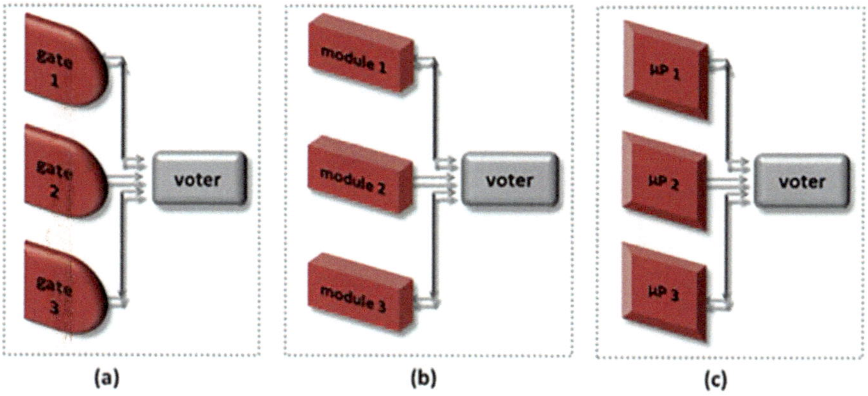

Fig. 3.7 Redundancy applied at different levels of abstraction: **a** three logic gates in a TMR at the logic or gate level of abstraction; **b** three memory modules in a TMR configuration at the circuit abstraction level; **c** three microprocessors in a TMR configuration at the chip level

TMR and *NMR* could be applied at different levels of abstraction triplicating logic gates, single memory cells, memory modules or complete microprocessors. Figure 3.7 displays how *TMR* can be applied at logic (a), circuit (b) and chip level (c).

TMR and *NMR* are typically employed in aerospace applications where the cost of failure is particularly high.

However, the higher reliability of these systems involves more than 200% increase in redundancy. Such an example is the *NASA* Space Shuttle on-board system, which is based on four computers with a majority voter (Sklaroff 1976).

3.4.2 Dynamic Redundancy

To reduce the extensive space, energy and performance overheads of *TMR* and *NMR* systems, numerous approaches have been developed. These approaches are usually based on dynamic redundancy, which implements error processing. This type of redundancy is similar to static redundancy with the main difference being the voter logic is replaced with a switch that is controlled by an error detection block.

At least one of the modules is working as the main module, whereas the rest of the modules or replicas can either be working in parallel (e.g. *DMR* with comparison) or can be turned off and used as spares (stand-by redundancy).

To avoid failures, after a fault has been detected, the system must be reconfigured. Detection, reconfiguration and recovery are required in order to prevent error propagation. Some examples of this type of redundancy are: pair and spare, duplex systems (*DMR* with comparison), backup sparing techniques etc.

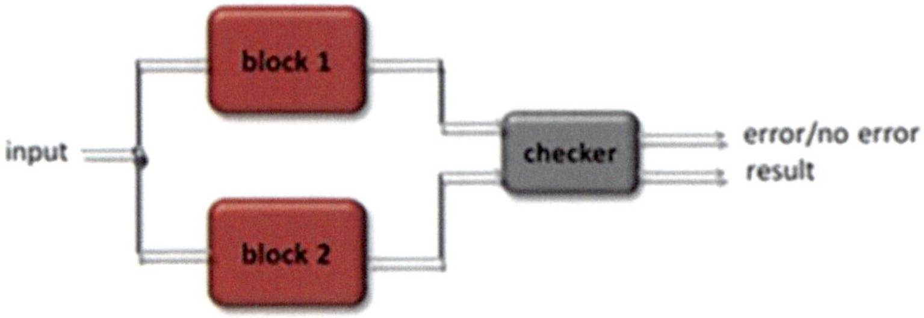

Fig. 3.8 Dual modular redundant (*DMR*) structure

3.4.2.1 Dual Modular Redundancy: HW(2*S*) + HW(δS)

By duplicating two components and adding a comparison structure, Dual modular redundancy (*DMR*), or duplex, a common system to detect errors (von Neumann 1956).

DMR uses two fully redundant units working in parallel and has been widely used in low level circuit implementations where a signal is duplicated as an input to two redundant and independent logic gates and it is transparently checked for errors.

Figure 3.8 depicts a *DMR* structure with a checker component. The checker logic compares the output of block 1 and block 2.

In the case of normal execution with no error, both blocks would produce the same output and a result would be delivered. On the other hand, in case of a mismatch between the two outputs of the blocks, the output of the checker would produce an error signal and no result will be given.

Therefore, in its simplest version, as the checker logic is unable to identify the incorrect unit, DMR, through output comparison, it will only provide error detection and will not provide error recovery capabilities on its own.

Additional mechanisms will be needed to provide error recovery so that if one of the units experiences an error, the surviving/correct unit can continue execution. Upon successful repair/recovery, *DMR* is fully restored.

3.4.2.2 Redundant Execution

A widespread and simple implementation of coarse-grained *DMR* is lock-stepping, or lock-step execution (Buckle and Highleyman 2003; McEvoy 1981; Sherman 2003). Here, the processor pipeline is duplicated and the clock is shared, comparing each instruction result before committing the results. This type of error detection is considered to perform at the macro-level since it is applied at the microprocessor's scope.

Lock-stepping is widely used in a number of commercial processor designs and can both detect and correct certain errors; e.g. *IBM G5* (Slegel et al. 1999) and Compaq Himalaya (Wood 1999).

Redundant threads are executed in multiple processors and every instruction result is compared. No instruction can be committed until its identical pair has also been completed and verified, hence involving considerable overhead.

3.4.2.3 Stand-by Redundancy

Standby redundancy, standby replacement, or standby sparing, is a well-known fault-tolerant design technique used as a failover mechanism (Avizienis 1976). In this case, some units are online and operational and one or more backup units serve as standby units.

Figures 3.9 and 3.10 present simple and multiple standby configurations. When a fault is detected in an online/active unit, a standby unit replaces the affected unit by using the selector (Fig. 3.9) or by using the 3-to-1-switch (Fig. 3.10).

There are three common forms of standby redundancy: hot, warm and cold. The type of application plays a key role in selecting the type of standby spare units. Figure 3.11 graphically describes the typical reconfiguration steps for hot, cold and warm backup spares.

When a spare unit is to be switched off, the selected spare is powered up and gets ready to become active. The reconfiguration process whereby a standby spare unit becomes operational is composed of:

1. Switching on the power and the bus connections
2. Powering up of the unit

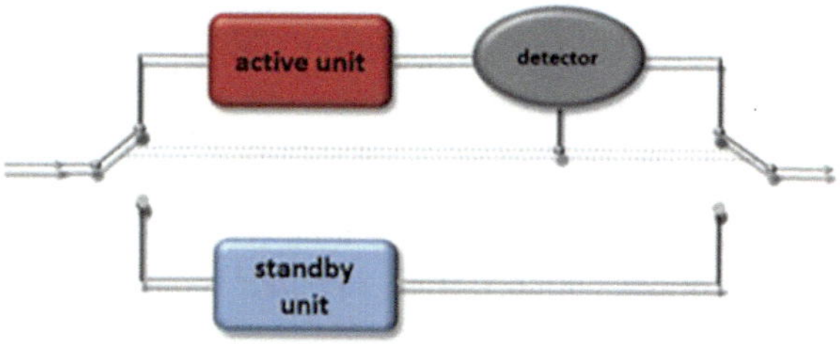

Fig. 3.9 Simple stand-by sparing configuration

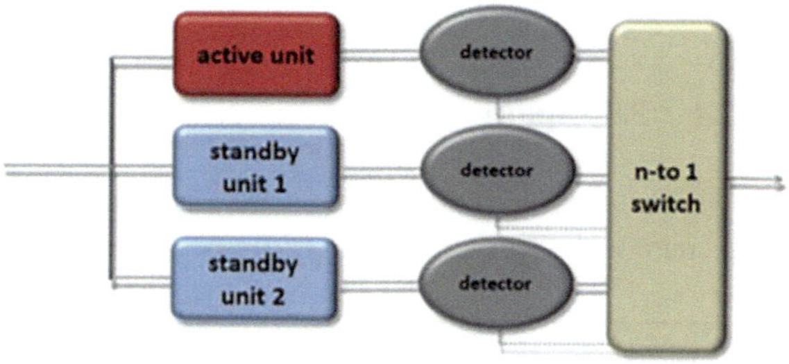

Fig. 3.10 Multiple stand-by spares with *n*-to-1 switch

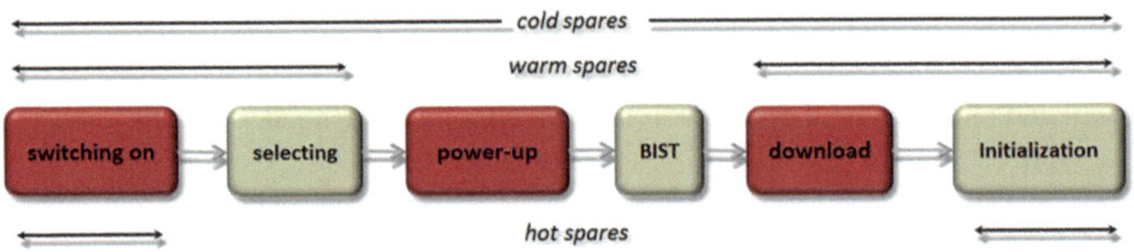

Fig. 3.11 Typical reconfiguration steps for back-up sparing

3. Running the *Built-In-Self-Test*[4] (*BIST*): Extensive testing is usually done after powering up to avoid replacing a faulty module with a faulty module before starting normal operation, e.g. memory tests of a spare module
4. Loading programs and data
5. Initializing the software if needed.

Hot standby spares (HSP) operate in synchrony with the operational units and are ready to take over whenever a fault is detected. *HSP* units reduce the mean time to recovery (*MTTR*) and therefore their use is suitable for applications that require short recovery time, that is, applications where the disruption of processing must be minimized.

Cold standby spares (CSP) remain unpowered and thus do not operate or consume any power until they need to replace an active unit. Since the restarting of the units is required, the use of *CSP* is best suited to remote operations where power is hard to come by, e.g. satellites and sensor systems. *CSP* units are also suitable for applications where short lapses in operation are acceptable and state data is not critical.

In addition, *CSP* are likely to have a lower failure rate than operational modules. However, the start-up delay required to switchover to a spare module is high since power up, *BIST* and initialization are needed. In particular, the time necessary for *BIST* depends upon the fault *coverage* and the complexity of the unit/module.

Warm standby spares (WSP) consist of a trade-off between the high-power consumption of *CSP* and the long recovery time of *HSP*. *WSP* units have time-dependent behaviour. Before and after replacing an operational unit, *WSP* presents different failure distributions.

The advantage of standby sparing for a system with n identical units is that a certain level of fault tolerance can be provided with $k < n$ spare modules.

3.4.2.4 Pair Spare

The *pair and spare* configurations are a combination of *DMR* with comparison and extra spare techniques. Figure 3.12 depicts a pair and spare configuration where two

[4] Built-In-Self-Tests (BISTs) are one of the common methods of testing circuits. BIST is a DFT technique that takes place on the same substrate as the device under test (DUT) within the system allowing them to perform self-testing (Stroud 2002).

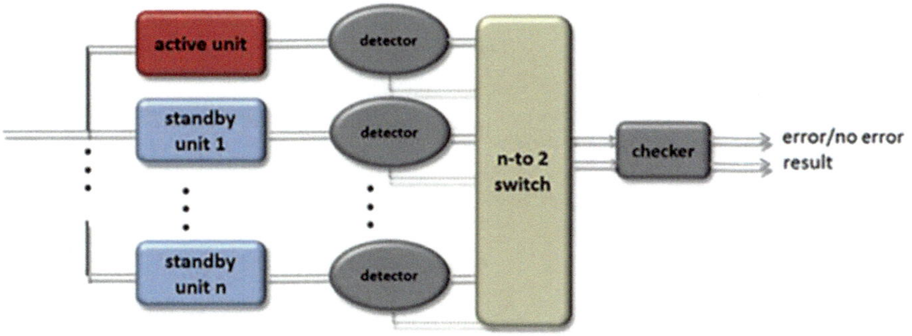

Fig. 3.12 Pair and spare configuration

units are always online and compared to each other, with any of the *n* spares being able to replace either of the operational units.

3.4.3 Hybrid Redundancy

By mixing fault masking, detection location and recovery, the advantages of static and dynamic redundancy can be combined (Johnson 1989). Hybrid approaches use Fault masking to prevent erroneous results from being activated. Fault detection, location and recovery are also employed in hybrid techniques to improve fault tolerance by removing errors.

A general approach is to back up the replicated modules with spares, e.g. a *TMR* configuration with a fault/disagreement detector, a voter and a reconfiguration unit (see Fig. 3.13).

In such a system, the triplicated operational modules are backed up with an additional pool of spares that can replace faulty modules (*TMR with spares*). The system will work as a basic *TMR* configuration until the disagreement detector determines that a faulty module exists.

One alternative approach towards fault detection is to feed the output of the majority voter back to the faulty detection unit whose job is to compare the output of the voter with the individual outputs of each operational module.

Any disagreement with a specific module's output would indicate that the module should be labelled as faulty and therefore replaced by a spare unit.

The reliability of the basic *TMR* system is retained as long as the pool of spares is not exhausted. Note that voting only occurs among the operational modules in the *TMR* core, masking faults and making sure that continuous correct service is delivered.

A variation to NMR with spares is the triple-duplex approach depicted in Fig. 3.14 that combines duplication with comparison and TMR. The use of passive redundancy

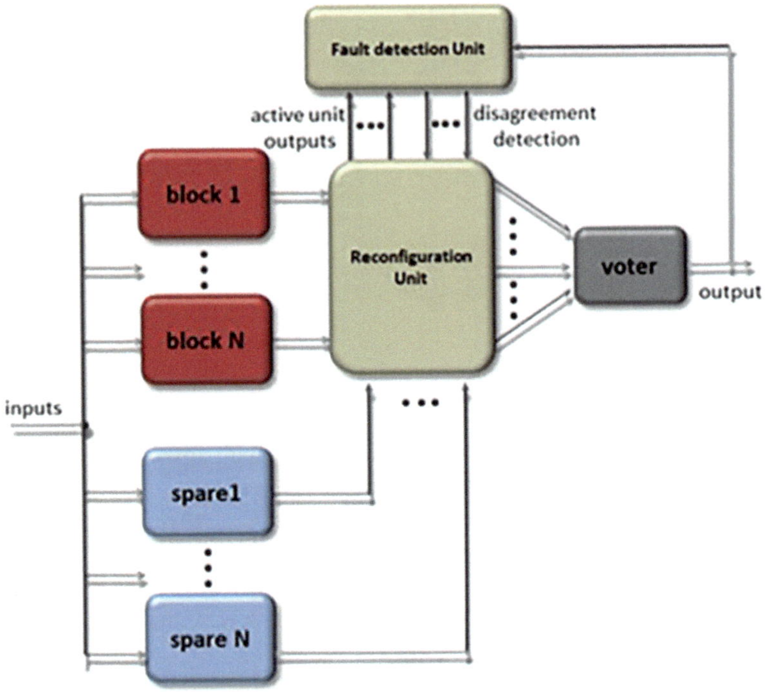

Fig. 3.13 Hybrid approach using TMR with spares

in the form of TMR allows potential faults to be masked and continuous correct service to be provided for a maximum of two faulty modules.

The use of DMR with comparison allows faults to be detected and faulty modules to be removed from the voting process and replaced by spares.

These options are simple but far more expensive in terms or real estate of the chip than traditional static techniques. Besides, as seen in Sect. 3.4.1.2, the reliability of TMR depends mostly on that of the voters.

Hence, if a fault takes place within a voter, an incorrect majority vote may be given to the output and propagated throughout the system, thus compromising the correctness of the system's service. In order to avoid such unreliability, voters can be designed to be capable of testing themselves online with regards to their own internal faults (Cazeaux et al. 2004; Metra et al. 1997, p. 97).

An extra form of hybrid scheme that allows error detection and correction, thus improving the reliability of a memory system (Fig. 3.15).

Any reading and writing operation is followed by a content check of a specific address in all three blocks. In case of a mismatch among these, a majority voting takes place whose result is then rewritten (via control unit) to the inputs of all elements using the same address.

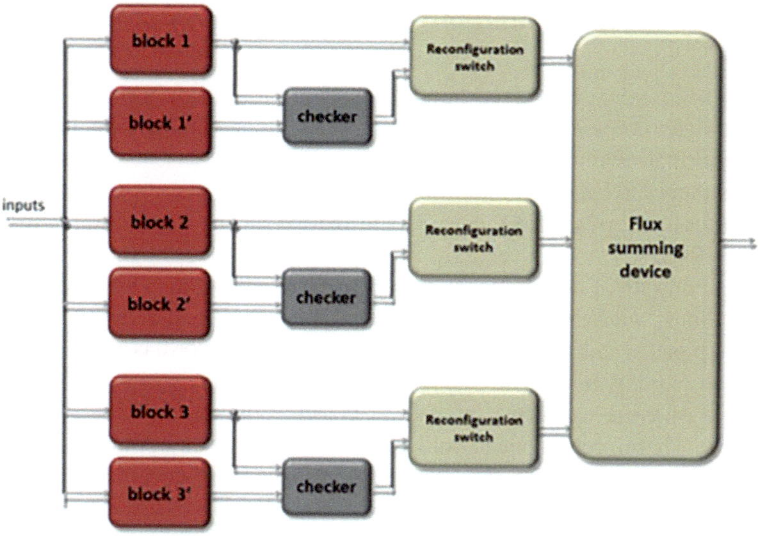

Fig. 3.14 A triple-duplex approach

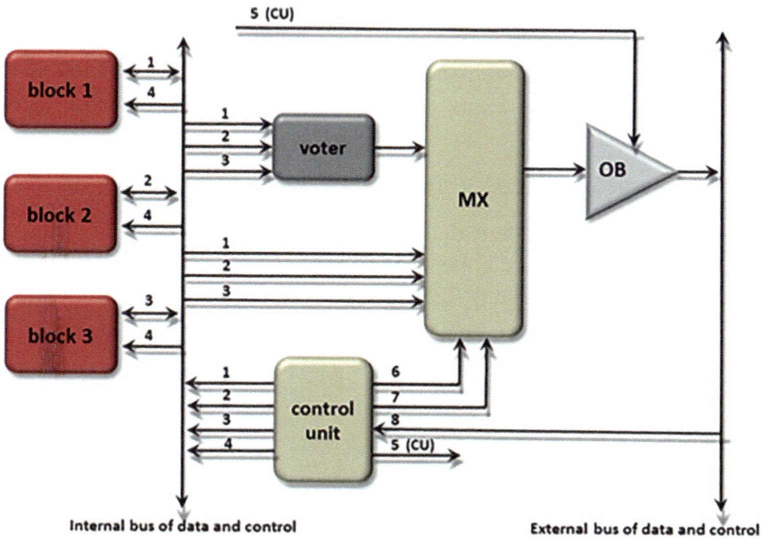

Fig. 3.15 Transient fault-tolerant TRAM (Schagaev and Buhanova 2001)

3.5 Information Redundancy

Information redundancy involves the addition of new information to existing information, often in compressed form, i.e. using more bits than needed. The most common form of information redundancy is *coding* (see Fig. 3.16).

Coding theory in hardware and software fault tolerance goes back a very long way and was initially motivated by the need to mitigate errors in information transmission (Shannon 1948).

Coding consists of adding check bits to the data allowing (1) the verification of data correctness and/or (2) the correction of erroneous data. Therefore, with coding an original piece of meaningful information, or *d-bit* data word, is encoded obtaining a *c-bit* code word, where $c > d$ (see Fig. 3.16).

Because of these extra bits, not all 2^c possible binary combinations are valid code words. Therefore, a code should be selected so that any potential error would transform the codeword, after decoding, into an invalid code word (non-codeword).

An important property of coding is separability. Two main approaches are possible:

1. Separable or systematic codes: the code word is formed by adding extra information (check bits) to the original data. A separable code has separable fields for data and check bits. Decoding this type of code is simple and consists of selecting the data bits and disregarding the check bits.
2. Non-separable or non-systematic codes: data and check bits are integrated together, requiring some extra processing and therefore incurring additional delays and overheads.

Important parameters for codes are:

1. The number of erroneous bits that can be detected
2. The number of those that can be corrected
3. The number of additional bits that are required

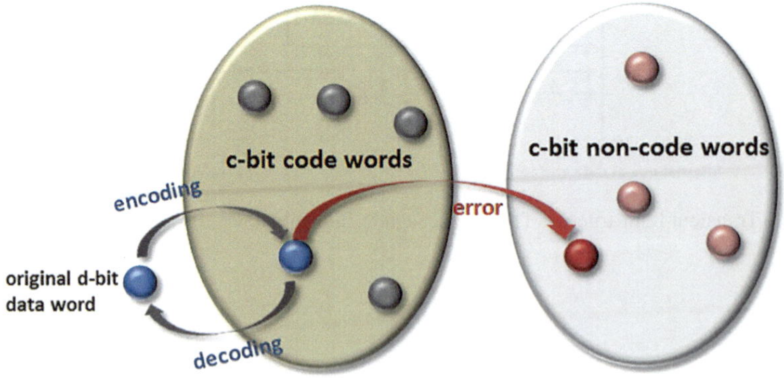

Fig. 3.16 Coding-encoding process of a *d*-bit word into a *c*-bit word

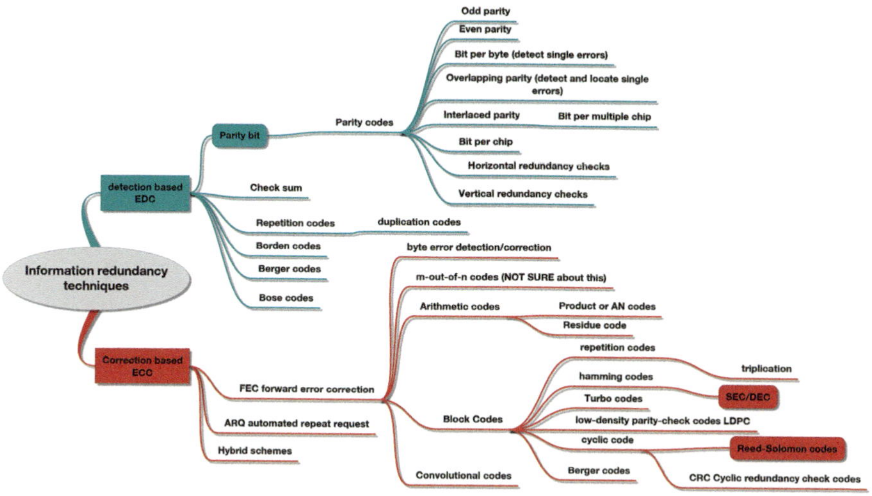

Fig. 3.17 Taxonomy of information redundancy coding techniques

4. The time needed to encode
5. The decoding time.

Information redundancy techniques make use of detection-based codes (EDC) or correction-based (*ECC*) codes. Figure 3.17 presents a taxonomy of coding techniques.

Examples of some of these techniques include:

1. Error detection and correction codes for cross-checking the contents of main memory, register files and cache,
2. Cross-checking of run-time control flow using signatures and
3. Algorithm-based checksums for cross-checking of the data values generated.

In general, information redundancy involves some space and computational overheads, thus requiring extra circuitry and is thus more commonly implemented in memory structures instead of in processor data-paths.

3.5.1 Error Detection Codes: EDC

Error detection codes have the ability to expose error(s) in a given data word based on the encoding–decoding principles discussed in Sect. 3.5. In general, error-detecting codes (*EDC*) present less overhead than error correcting codes (*ECC*) since they do not have correction capabilities.

The simplest *EDC* are parity codes, which involve the addition of extra bits or parity bits depicts a basic scheme of memory with parity checking. Figure 3.18 depicts a basic scheme of memory with parity checking.

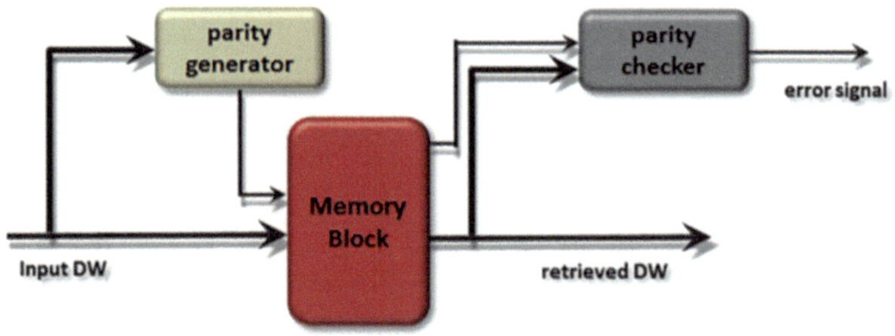

Fig. 3.18 Coding-encoding in memory block with parity checking

Before storing a word in the memory block a parity generator computes the parity bits based on the bits of the input data word (*DW*). A parity bit is an extra bit added to a group of source bits (*DWs*) in order to ensure that the outcome or coded word has an even (in the case of even parity) or odd (in case of odd parity) number of bits set to 1.

When a memory block is read, the parity checker compares the computed and the stored parity bits, setting the error signal consequently. If both, computed and stored parity bits, match then the error signal would indicate a correct output; otherwise, the error signal would indicate that the retrieved *DW* is incorrect. Note that for n bits of data there are 2^n possible *DWs*. Adding one parity bit would allow 2^{n+1} possible *DWs*.

Among these possible *DWs* there are $\frac{2^{n+1}}{2}$ possible *DWs* with an odd number of 1s and $\frac{2^{n+1}}{2}$ possible *DWs* with an even number of 1s.

In the case of odd parity, only the *DWs* with an odd number of 1s are valid code words (*CWs*). In the presence of single bit flip (error) an odd parity *CW* would change into an even parity *CW* and therefore the parity checker will detect the error. Nonetheless, it will not know which bit has been flipped. This simple configuration can be used to detect single or any odd number of errors in the retrieved *DW*. However, an even number of flipped bits would make the parity bit of the *CW* appear to be correct, although the data is incorrect. With single parity, double errors and even number of errors would remain undetected.

3.5.2 *Error Correction Codes: ECC*

Codes more powerful than parity can be created by adding more check bits to the original data. The size of the data to be protected will determine the number of check bits needed. Using this basic principle, error correction codes have the ability to detect errors and reconstruct the original error-free data. These can generally be realized in three different manners (see Fig. 3.17):

1. *Backward Error Correction (BEC)* sometimes referred to as Automatic repeat request (ARQ): combines an error detection technique (error detection encoding prior to transmission) with retransmission request of erroneous data. BEC requires simpler decoding infrastructure than FEC but frequent retransmissions would significantly compromise performance in high data rate transmissions.
2. *Forward Error Correction (FEC)* or Channel Coding: With this approach, errors are both detected and corrected at the receiver's end. Thus, it involves error-correcting encoding prior to transmission without retransmission of the original information. FEC requires more complex decoding infrastructure than BEC, but it is suitable for high data rate applications.
3. *Hybrid Automatic Repeat Request (HARQ)*: BEC and FEC are combined. E.g.: a scheme where minor errors are corrected without retransmission (FEC) and major errors are corrected via retransmission (BEC).

An overview of popular *FEC* schemes employed in fault tolerant design of embedded systems follows. Figure 3.19 shows a scheme proposed in Schagaev et al. (1992, 1993). *ECC* memory scheme that applies to any of the following codes, including calculation, checking and correcting logic.

When data is written into the data row specified by the address signals, the *ECC* encoding logic generates the parity checks (as specified by the code) and introduces them into the *ECC* part of the memory.

When the DW is read from the memory the parity bits would allow missing data to be reconstructed in the case of an error being detected.

3.5.2.1 SEC-DED[5]: Hamming and Hsiao: HW(δI)

The most common *ECCs* are based on Hamming (1950) or Hsiao (1970). These two separable code families introduce the concept of overlapping parity by which every data bit has a part in adjusting the value of several parity bits. These codes can correct single bit errors in a given word, can detect double bit errors, are relatively fast decoding and have moderate redundancy. Hamming codes are a family of perfect codes[6] that generalize the original Hamming (7,4)-code (Hamming 1950). A minimum distance *d* means that it takes *d* bit changes to move from one valid codeword to the other.

Extended Hamming code, sometimes generalized as *SEC-DED* (single error correction and double error detection), is an example of this type of code. In *SEC-DED*, an extra parity bit is added, achieving a distance of four instead of the three (as in the original Hamming).

---

[5] SEC-DED: Single error correction and double error detection.

[6] A Hamming code is perfect in the sense that it can achieve the highest possible rate for codes with a given block length and minimum distance of three (Moon 2005).

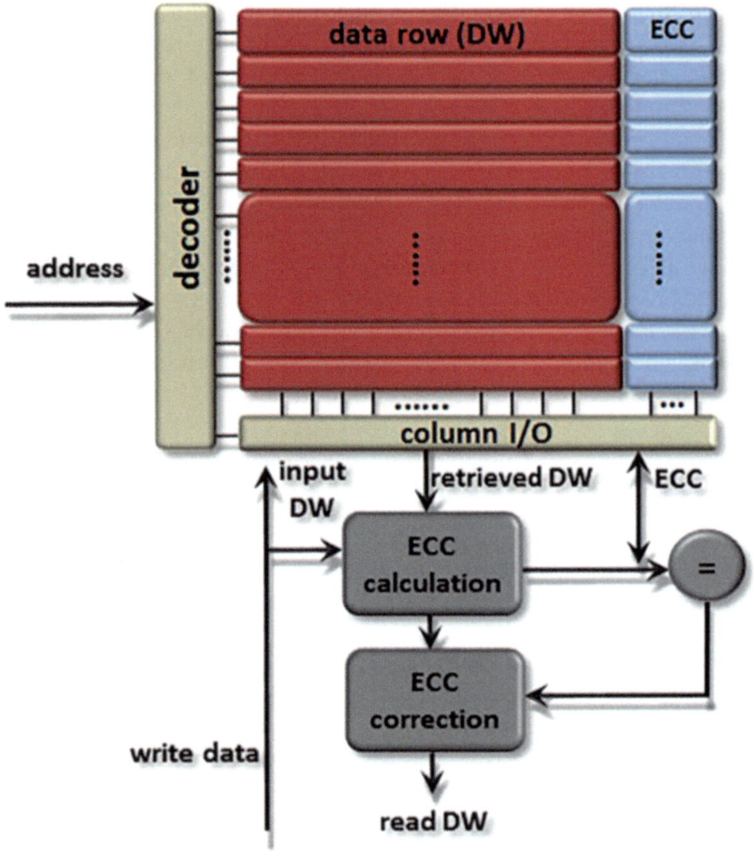

Fig. 3.19 Basic ECC memory scheme including calculation, checking and correcting

The extra parity bit allows the decoder to distinguish between two possible situations:

1. When at most one-bit flip has occurred and
2. When two-bit flips have taken place.

In contrast with *Hamming (7,4)*, *SEC-DED* provides single-bit-error correction and simultaneous double-bit-error detection.

Compared to Hamming codes, Hsiao codes (Hsiao 1970) provide improvements in speed, reliability and calculation cost as well as checking and correcting logic.

However, in situations that demand higher reliability requirements than those provided by *SEC-DED*, more complex codes are required.

SEC-DED Limitations and Alternative Techniques

The main limitation of *SEC-DED* codes is that triple-bit errors may not only remain undetected but it may also be mis-corrected as if they were single-bit-errors (Hsiao 1970). The probability of this type of mis-correction for 32 bit data words is around 60% or more.

Multiple errors are usually taking place in adjacent memory locations, therefore increasing the chances of having multiple bit errors in a given word (Bentoutou and Djaifri 2008; Boatella et al. 2009). These are called *burst errors*,[7] errors that are highly correlated.

If a specific memory cell has an error, it's likely that adjacent cells may also be corrupted by the same event that triggered the error in the first place. These are sometimes referred to as *spatial multi-bit errors* (Mukherjee et al. 2004). In contrast, *temporal multi-bit errors* are errors that take place when two different cells of the same word are affected by different events (Mukherjee et al. 2004).

An important risk for *SEC-DED* schemes is that if a specific memory word is not accessed for a long period of time, the chance of accumulating errors increases (temporal multi-bit errors).

One method to avoid these is the use of *memory scrubbing* (Mukherjee et al. 2004; Saleh et al. 1990; Weaver et al. 2004), by which every memory location is read periodically. This may be implemented by having a hardware controller that, during idle periods, reads every memory location searching for errors and correcting any single error found during the process, thus reducing the chance of detected (*DRE*[8] and *DUE*[9]) and undetected errors (e.g. *SDC*[10]).

Scrubbing does, however, impose additional SW and/or HW overheads depending on the implementation. In current architectures with high memory bandwidths, HW scrubbing is preferred due to its lower timing overhead. In combination with *SEC*, scrubbing is effective against single-bit- and temporal multi-bit errors but not against spatial multi-bit errors.

To avoid this problem of spatial multi-bit errors, *memory interleaving* (Haraszti 2000; Reviriego et al. 2007, 2010) is commonly used in conjunction with *ECC*, ensuring that cells that are physically closely located belong to different logic words. That is, cells that belong to the same logical word are physically apart.

---

[7] Also called cluster of errors.

[8] A DRE is a detected recoverable error, a benign type of error since recovery of the normal operation by fault tolerant techniques is possible (Kadayif et al. 2010; Weaver et al. 2004).

[9] A DUE is a detected unrecoverable error. DUE take place when fault tolerant techniques are able to discover and/or report an error, from which recovery is not possible (Kadayif et al. 2010; Weaver et al. 2004).

[10] SDC stands for Silent data corruption. A SDC take place when an error is undetected and causes data corruption (SDC). In this case, the corrupted data could go unnoticed making this type of error benign, or could result in a visible error and/or catastrophic failure such as crashing a computer system (Constantinescu et al. 2008; Kadayif et al. 2010; Weaver et al. 2004).

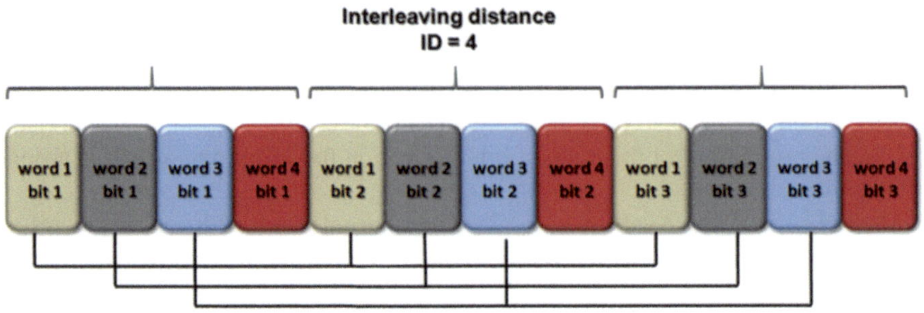

Fig. 3.20 Memory interleaving of four 3-bit words with a 4 interleaving distance

Figure 3.20 illustrates an example of memory interleaving in a four *3-bit* memory word. This type of memory distributes logical data into a non-continuous arrangement. More columns than the number of bits of a single word are added, and the corresponding columns for each word are interleaved. In this way, *burst errors* are distributed over a number of words each suffering only one single bit error.

Any *4-bit-upset* affecting adjacent memory cells would cause four single-bit errors in separate words, which can be easily corrected by *SEC-DEC*.

A shortcoming of interleaving is that high interleaving distances (*ID*) involve more complex designs and thus higher area and latency overheads (Baeg et al. 2009; Reviriego et al. 2010). Ideally, the ID should be selected as the maximum expected *MCU* size so that all upsets in a burst error would occur in different logical words.

3.5.2.2 Complex Codes

EDAC implementations based on Hamming codes are the easiest to implement but only provide single error correction (Hentschke et al. 2002). There are alternatives to *SEC-DED* like Bose–Chaudhuri–Hocquenghem (BCH) (Bose and Ray-Chaudhuri 1960) and Reed–Solomon (RS) codes (Reed and Solomon 1960) based on finite-field arithmetic that can correct multiple faults. Table 3.4 shows a comparison of the main error correction techniques in memories.

BCH codes are able to correct a given number of bits at any position, whereas *RS* codes group the bits in blocks in order to correct them. *RS* based codes provide a more robust error correction capability but uses a large amount of system resources.[11] The *RS* decoding process has several stages to get the location of the error and correct it. Implementations of *RS* codes can be found in Neuberger et al. (2003, 2005). Although the *RS* algorithm can be simplified (Neuberger et al. 2003) the main disadvantage of these two codes is having complex and iterative algorithms.

[11] DEC-TED implementations are expensive from both area-penalty and computational-complexity points of view.

Table 3.4 ECC-TMR comparison

Characteristic	Hamming (SEC-DED)	TMR	RS (DEC-TED)	BCH (DEC-TED)
Area	Small overhead to implement Varies depending on the number of bits (7–32%)	Extra 200% plus the voting and correcting logic Number of voters is proportional to the number of units	Varies depending on the number of bits (13–75%)	Varies depending on the number of bits (13–75%)
Performance	It can be affected by the coder-decoder functions Proportionally dependent on number of bits to be corrected	High performance. Voter is the only source of delay, hence, almost constant delay	Lower performance than *BCH* and much lower compared to Hamming and *TMR*	Higher performance than RS but much lower than Hamming or *TMR*
Error correction	Limited capabilities: it corrects only one single incorrect bit per word	Corrects up to *n* errors in an *n*-bit word as long as the errors are located in a distinct position/unit	Can handle multiple errors; efficient for correlated errors (e.g. burst)	Can handle multiple errors; efficient for uncorrelated errors (e.g. random errors)
Implementation	Binary based Simple to implement	Simple to implement	Symbol based Complex to decode and implement	Binary based Complex but simpler to decode and implement than *RS*

As with hamming based *SEC-DED*, more complex codes can be implemented based on RS and BCH algorithms. Some examples are *SNC-DND*[12] (Chen and Hsiao 1984) and *DEC-TED*[13] codes (Lin and Costello 1983).

Table 3.5 is an overhead comparison of various *EDAC* schemes: Single parity *EDC*, Hamming *SEC-DED, SNC-DND* and *DEC-TED.* Complex errors increase the overhead rapidly as correction capability is increased (Kim et al. 2007). Note also that for any given technique, as the data size increases, the relative overhead of a given scheme decreases (Table 3.5). Note that the calculation of overheads is just the number of check bits divided by the number of data bits and does not include the extra overheads (e.g. I/O and checkers).

In addition to the area penalty, as the correction capability increases, timing overheads also increase. Results on 64 kb *SRAM* developed in 90 nm processes show that the implementation of a *DEC-TED* encoder involves a latency penalty of 80–85% as compared to *SEC-DED* (Naseer et al. 2006). Schemes based on information redundancy can also be applied at different levels. For instance, parity codes can be applied

[12] SNC-DND: single nibble error correcting, double nibble error detecting.

[13] DEC-TED: double bit error correcting, triple bit error detecting.

Table 3.5 EDC-ECC storage array overheads, based on Slayman (2005)

Data bits	Single parity		SEC-DED		SNC-DND		DEC-TED	
	Check bits	Overheads (%)	Check bits	Overheads (%)	Check bits	Overheads (%)	Check bits	Overheads (%)
16	1	6	6	32	12	75	11	69
32	1	3	7	22	12	38	13	41
64	1	2	8	13	14	22	15	23
128	1	1	9	7	16	13	17	13

to registers, cache and internal memory whereas *SEC-DED* can be implemented in external memory, etc.

As all these are more complex codes than *SEC-DED* let alone single parity codes they produce higher overheads as the correction capability increases (Kim et al. 2007) and are thus not suitable for areas of real-time systems that demand high possessing performance.

3.6 Time Redundancy

Figure 3.21 shows a list of the most relevant techniques based on time redundancy, which are fully described in Sects. 3.6.2, 3.6.3, 3.6.4, 3.6.5 and 3.6.6.

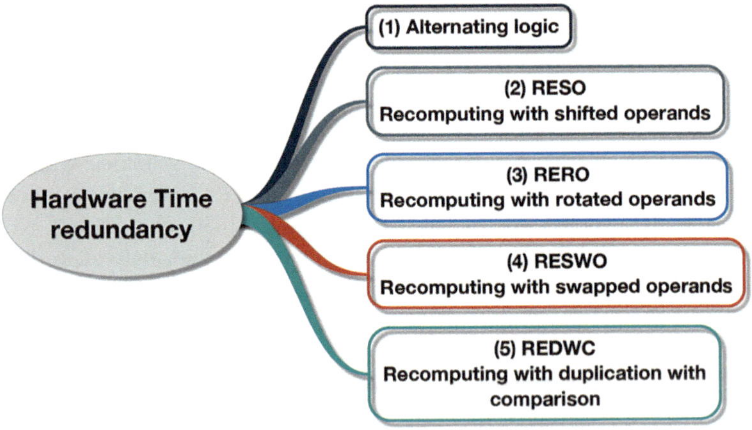

Fig. 3.21 Taxonomy of time redundancy techniques

3.6.1 Concurrent Error Detection: Basics of Time Redundancy

The main problem with the structural and information redundancy types reviewed is the penalty imposed in the form of extra hardware. At the expense of using additional time, *FT* techniques based on time redundancy (*TR*) aim to reduce the amount of hardware required for the implementation.

Time redundancy techniques involve the re-execution of code using the same piece of hardware and comparing the two execution results to determine if a fault has occurred. This approach was commonly used in the past and is effective in detecting errors resulting from transient faults.

The main problem with the space and information redundancy types reviewed is the penalty imposed in the form of extra hardware. At the expense of using additional time, *FT* techniques based on time redundancy (*TR*) aim to reduce the amount of hardware required for the implementation. Time redundancy techniques involve the re-execution of code using the same piece of hardware and comparing the two execution results to determine if a fault has occurred. This approach was commonly used in the past and is effective in detecting errors resulting from transient faults.

Figure 3.22 shows the basic transient fault detection mechanism based on re-execution. With this technique two or more different computations are performed at different times t_0, $t_0 + \Delta t$, and $t_0 + n\Delta t$ given $n > 1$.

The result of a given computation is stored in the corresponding register and then compared to the results obtained from the previous computation(s). If the re-execution is performed twice and a disagreement exists, then transient errors can be detected. This type of technique was used in the past, but on its own, and did not provide protection against errors due to permanent faults.

However, the executions can be performed again to check if the discrepancy remains or not. This is useful in order to distinguish between permanent and transient faults. If after re-execution the fault disappears, it is assumed to be transient.

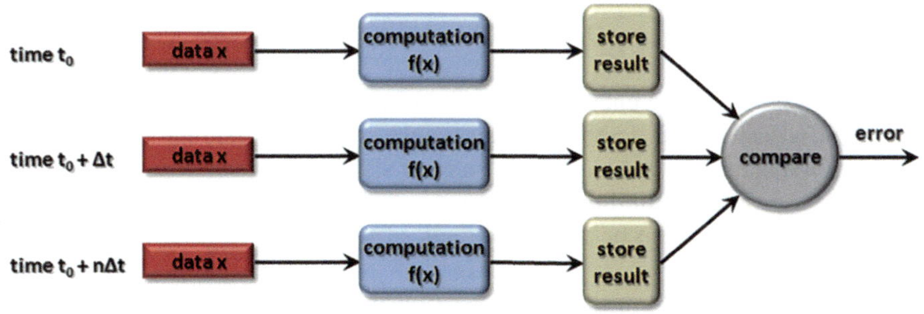

Fig. 3.22 Transient fault detection mechanism based on redundant execution

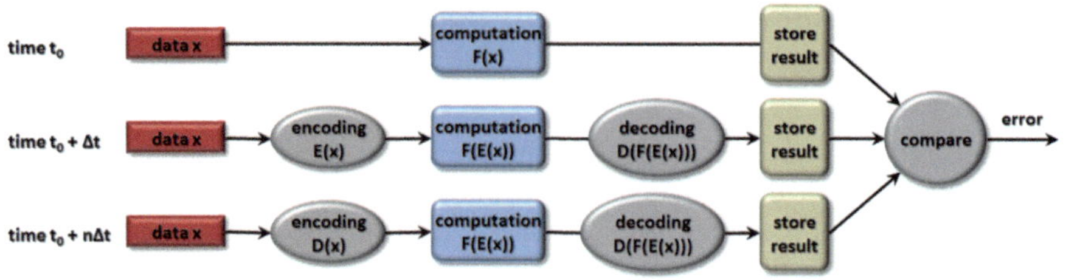

Fig. 3.23 Transient and permanent fault detection mechanism based on redundant execution

The hardware resource affected by a transient fault is still usable. On the other hand, if after re-execution the problem persists, the fault is assumed to be permanent and reconfiguration of the specific hardware resource is necessary.

Modern *FT* techniques based on time redundancy can detect permanent faults as shown Fig. 3.23. In this case, during the first computation, the results obtained are simply stored in a register. Then, prior to the next computation(s) a specific type of encoding is performed on the operands. After the relevant computation(s) take(s) place on the encoded operands, the results of all computations are then decoded and compared.

Given that x is the input data, $E(x)$ is the data decoding, $F(x)$ is the functional computation, $F(E(x))$ is the functional computation of the decoding data and $D(E(f(x)))$ is the decoding of the encoded data after computation, and assuming that the functional block is free of permanent faults and assuming that:

Equation 3.8 Encoding–decoding relationship

$$D(E(x)) = x \quad \forall x \tag{3.8}$$

Then, the following relation can be stated:

Equation 3.9 Relationship among encoding, decoding and functional computation

$$D(F(E(x))) = F(x) \quad \forall x \tag{3.9}$$

If the decoder and the encoding process are carefully selected so that a failure in x would affect $F(x)$ differently than it would affect $F(E(x))$ then if $\Delta t > 0$ the comparison mechanism would produce an error signal.

The main problem with time redundancy techniques is that if the system's data is corrupted by a transient or permanent fault, it will be difficult to repeat a given computation.

The critical part of these techniques is assuring that the data is correct and identical before each one of the redundant computations takes place.

The leading concurrent error detection (*CED*) techniques based on time redundancy are *alternating logic*, *recomputing with shifted operands* (*RESO*), *recomputing with rotated operands* (*RERO*), *recomputing with* swapped *operands* (*RESWO*) and *recomputing with comparison* (*REDWC*).

All these techniques work as specified in Fig. 3.23. The main difference among them is the type of encoding and decoding used.

3.6.1.1 Self-duality

Self-duality is a property required for certain circuit's functions in order to implement specific error detection techniques based on time redundancy. A function is said to be self-dual if it satisfies the property:

Equation 3.10 Property of self-duality

$$f\left(x_{1\_}, x_{2\_}, \ldots, x_{3\_}\right) = f\_(x_1, x_2, \ldots, x_3) \quad \forall x \tag{3.10}$$

where x_1, x_2, x_3, …, x_n is the set of inputs to the circuit, $x_{1\_}$, $x_{2\_}$, …, $x_{n\_}$ the set of complemented inputs, $f()$ the output and $f\_()$ the complemented output.

By letting C be a function that complements each bit of a given vector:

Equation 3.11 Property of self-duality

$$C\left(x_{1\_}, x_{2\_}, \ldots, x_{3\_}\right) = (x_1, x_2, \ldots, x_3) \quad \forall x \tag{3.11}$$

It becomes clear that:

Equation 3.12 Complementary function

$$\begin{aligned} &C^{-1} = C \\ &C^{-1}(f(c(x))) = f(x) \end{aligned} \tag{3.12}$$

Resulting in:

Equation 3.13 Complementary function and self-duality

$$C(f(x)) = f(C(x)) \tag{3.13}$$

There are several problems that must be considered when designing a fault tolerant technique using time redundancy. A function C that satisfies the previous property must firstly be determined. Finding a C may not guarantee the desired level of error detection since different circuit implementations based on different C can have different coverage.

Complexity is also an important issue. In the case where the hardware required to implement the coding and decoding functions based on C and C-1 is similar to that of implementing $f(x)$, then structural redundancy becomes the more effective

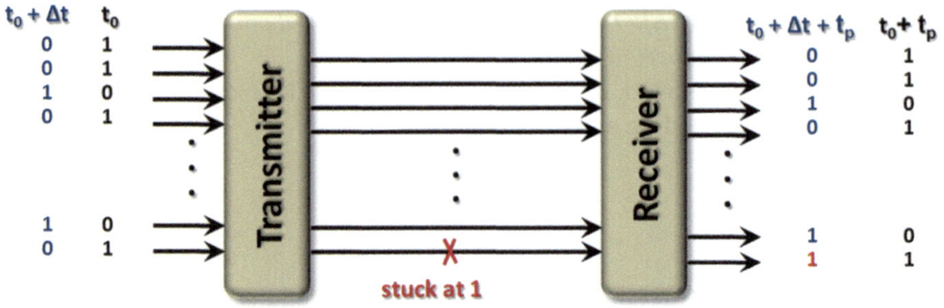

Fig. 3.24 Time redundancy techniques based on alternative logic

choice. In short, the aim of a cost-effective design should be finding a function C that provides a good trade-off between high coverage and low complexity.

3.6.2 Alternating Logic

An example of encoding/decoding function is the complementation operation used in *alternating logic* (Reynolds and Metze 1978) and successfully applied to permanent fault detection of data transmission and digital systems.

As shown in Fig. 3.24 the data computed at time t_0 is then complemented and transmitted at time $t_0 + \Delta t$. In the case of a stuck line (either at 0 or at 1) the two computations will generate data that are not complement of each other and therefore the error signal will become enabled after comparison.

In this example, the last communication line is *stuck at 1*[14] and, therefore, complement and data would both become 1, which is not an alternate output and therefore a fault is detected. In order to implement error detection in this coding the circuit function must have the property of *self-duality* otherwise extra input bits would be required. For certain circuits, 100% area overhead may be required for certain error detection circuits in addition to time redundancy (Carter and Schneider 1968; Johnson et al. 1988; Johnson 1989; Woodard and Metze 1978).

The key for fault detection is to determine that at least one input vector exists for which the fault will not result in alternated outputs.

Although any single *stuck-at fault* can be detected by this technique, extra redundancy and hardware modifications are required to create self-dual functions from non-self-dual ones. Any non-self-dual function of x variables can be converted into an $x + 1$ variable function that is self-dual and can thus be implemented with an alternating logic circuit.

---

[14] A stuck-at fault is a particular fault model used to represent a manufacturing defect within an integrated circuit. Depending on the effect of the fault, a suck-at fault can be stuck either at a logical value of 0 (stuck-at 0) or 1 (stuck-at 1).

3.6.3 Recomputing with Shifted Operands (RESO)

Recomputing with shifted operands is a logic level concurrent error detection technique based on time redundancy developed by Patel and Fung (1982). *RESO* can be applied to certain problems in which shifting the inputs forms a complementing function that produces a known relationship in the outputs. It has been originally used for arithmetic and logic units.

The error detection capability of *RESO* depends on the number of shift operations. The generalized version is *RESO-k* and it refers to shifting by k bits.

Figure 3.25 shows a schematic of a concurrent error detection mechanism on an *ALU* using *RESO*. The operands a and b undergo a normal *ALU* operation $f(a, b)$ during the first computation at time t_0 and the result is stored in a register.

During the second computation at time $t_0 + \Delta t$, before entering the *ALU* the operands are shifted left by k bits and the result of the *ALU* operation is right-shifted and finally compared to the ones previously stored in the register.

In such operations, left and right shifting can also be denoted as $E(x)$ and $D(x)$ (or $E^1(x)$). Therefore, if the equivalent notation for the re-computation is $E^{-1}[f(E(a, b))]$ it should be equal to the first computation $f(a, b)$. If the results are identical, the output of the computation will be $f(a, b)$.

However, if there is a discrepancy, an error signal will be generated. When an n-bit operand is shifted left by k-bit(s), its leftmost k bit(s) move out and the right

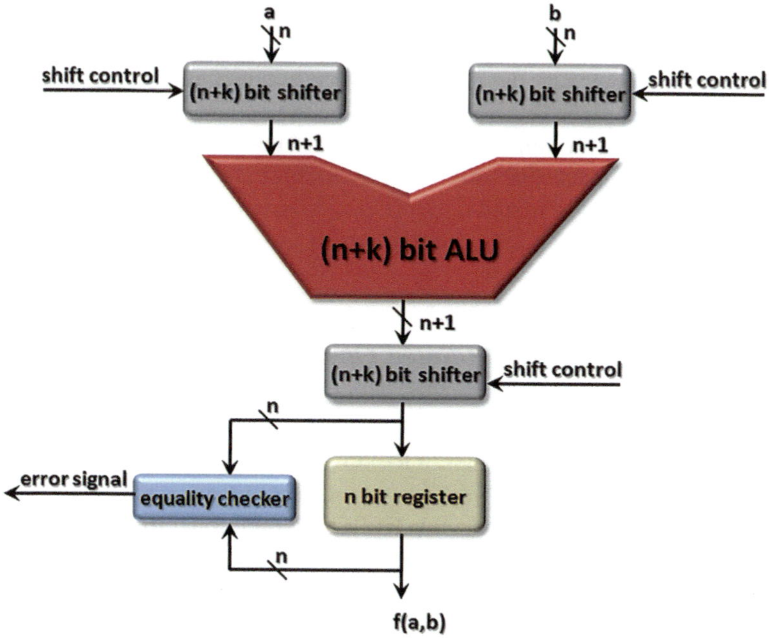

Fig. 3.25 ALU concurrent error detection using recomputing with shifted operands (*RESO-k*)

most k-bit(s) become zero. This may lead to an incorrect result of $f(a, b)$ since k essential bit(s) are removed whenever shifted left.

As with alternating logic, extra redundancy is needed as an $(n + k)$ shifters need to be implemented. Furthermore, a bigger $(n + k)$ bits length *ALU* is needed and therefore the re-computation takes $(n + k)$ bit operations rather than the original n-bit ones.

Furthermore, a totally self-checking equality checker is required for the comparison process and error signalling. Additionally, parity codes can also be used to detect error in the shifter logic.

Note that the fault coverage capability of *RESO* depends on the number of shifts. *RESO-1* can detect all single bit-slice errors in an *ALU* for all bitwise operations, including *AND*, *OR*, *NOT*, *NOR* and *XOR*. As k becomes larger, an increase of space and time complexity is entailed which in turn increase the probability of error.

Consider an *ALU* with an n-bit shifter and a *RERO-2* implementation and an operand a equal to *11010*. After the *11010* is being shifted left by two bits, it will have the two *MSBs* shifted out, thus becoming *01000*. As a consequence, the result of the calculation $f(a, b)$ will probably be incorrect.

If the shifter is replaced by an $(n + k)$*-bit* shifter with $k = 2$ in this particular case (*RESO*-2), then the operand a after the shifting operation will be equal to *1101000*, thus keeping the *MSBs* and ensuring the correct result $f(a, b)$. Note that during the first computation, *k-zero MSBs* are added to each of the operands.

This is one way of detecting errors using *RESO*. Alternatively, as before, during the first computation at time t_0, the operands a and b undergo a normal *ALU* operation $f(a, b)$ but the results are now left-shifted before being stored in the register.

In the second computation at time $t_0 + \Delta t$, the operands are also left-shifted by k bits, but in this new way, the results are directly compared with the ones in the register (there is not right-shifting performed on the operands).

The penalty paid for implementing *RESO* is that every component must be extended to accommodate the shifting. For instance, to implement *RESO-1* on a *32 bit ALU*, the main system and the shifters are required to be *33* bits, whereas the storage registers and the equality checker must be *34* bits.

3.6.4 Recomputing with Rotated Operands (RERO)

Recomputing with rotated operands (Li and Swartzlander 1992) is another technique designed to overcome the limitations of RESO. RERO-k has similar time redundancy characteristics to RESO but with different structural redundancy demands.

Figure 3.26 displays an ALU with RERO-k-based concurrent error detection. RESO-k requires an $(n + k)$-bit rotator and an $(n + k)$-bit ALU whilst RERO-k only requires an $(n + 1)$-bit rotator and an $(n + 1)$-bit ALU.

During the first computation, the $(n + 1)$ bit rotators do not rotate the operands, thus the input and output of the rotators are identical. Both operands undergo a regular ALU operation whose result is stored in a register.

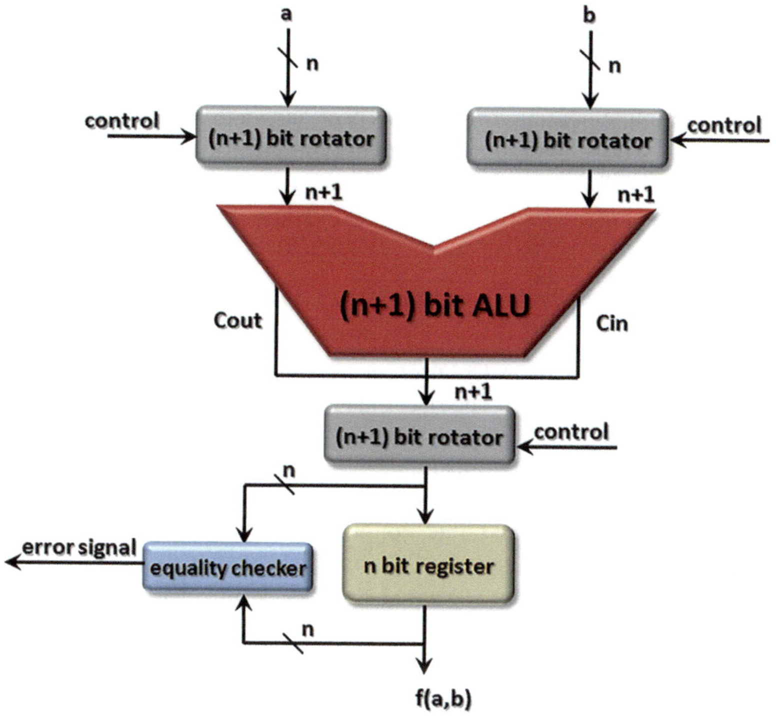

Fig. 3.26 ALU concurrent error detection using recomputing with rotated operands

During the second computations, the first two rotators perform a k-bit(s) right-rotation of the input operands before they enter the ALU. Next, the result is rotated left and compared to the previous result from the first computation. If the results are identical, the output of the computation will be $f(a, b)$.

However, if there is a discrepancy, an error signal will be generated. With regards to error coverage, a RERO-k implementation with an n-bit ALU can detect:

1. (k *mode* n) consecutive errors for bit-wise logical operations.
2. $(k - 1)$ consecutive errors in a ripple carry adder for arithmetic operations.

3.6.5 Recomputing with Swapped Operands (RESWO)

Recomputing with swapped operands (Hana and Johnson 1986) is an extension of the *RESO* technique that tries to detect errors by alternating the position of the operands. *RESWO* implementation is very intuitive but with limited applications such as addition, multiplication and Boolean functions but not division or subtraction operations.

The first computation at time t_0 is performed on unmodified operands. During recomputation, at time $t_0 + \Delta t$, the operands are first split into two halves, upper and lower, then swapped before calculation and finally swapped back after it.

The logic for implementing *RESWO* has been shown to be less complex and less expensive than in *RESO*, in particular when the complexity of individual modules is high (Shedletsky 1978).

3.6.6 Recomputing with Comparison (REDWC)

Recomputing with comparison (Johnson et al. 1988) uses a combination of both hardware and time redundancy. The operands *a* and *b* of an *n*-bit operation are split into two halves and computed by two virtually divided devices (*n*/2-bit size) twice. In a first time slot, the least significant *n*/2-bits (lower halves) of the operands and their duplicates are carried out and then their results compared.

Upon completion, in a second time slot, the same operation is repeated for most significant *n*/2-bits (the upper halves) of the operands. As long as the separate halves do not become faulty in the same way and at the same time, REDWC can detect all single faults.

3.7 Comparison of Main Redundancy Schemes

In general, the addition of correction capabilities to a mechanism involves extra area and/or time overheads. Table 3.6 compares the capabilities, timing and area overheads of structural- and time-based *FT* mechanisms.

TR techniques involve low area overheads at the cost of extra timing. Note that some of these techniques, such as *RESO*, can provide correction capabilities by performing more than two computations. In contrast, at the cost of area penalties, *SR* techniques can provide detection and correction with very little timing overheads.

Error detection codes, such as parity coding involve low complexity and low overheads but have limited detection abilities and are not able to detect multi-bit errors.

Although information redundancy schemes can be feasible to correct single and double errors in high-capacity memories (Paul et al. 2011), for $n > 2$, *n*-bit correction circuitry demands considerable area, energy and timing overheads, especially in low capacity memories.

For instance, in an *8-bit ECC* scheme integrated to a *64 kb* SRAM the area overhead can be more than *80%* (Kim et al. 2007). Application of Hamming *SEC-DED* codes to 16 M-bit *DRAM* chips has a 10% access time penalty on a *16 M*-bit *DRAM* (Arimoto et al. 1990; Furutani et al. 1990).

For an experimental *1 M*-bit *DRAM* cache, applying a *SEC-DED* code imposes up to *15%* access time overhead (Asakura et al. 1990). The area penalty is even greater

Table 3.6 Comparison of structural, time based FT mechanisms

Scheme	Structural redundancy (%)	Time redundancy (%)	Detection (D) and correction (C)
Time redundancy based			
Alternating logic	$\approx$ 0–100	> 100	D
RESO	$\approx$ 0–93	> 100	D
RESO	$\approx$ 0–93	> 200	DC
RERO	$\approx$ 0–93	> 100	D
RESWO	$\approx$ 0–77	0–100	D
REDWC	$\approx$ 0–90	0–100	D
Structural redundancy based			
DWC	> 100	$\approx$ 0–17	D
TMR	> 202	$\approx$ 0–17	DC
TMR with triplicated voter	> 208	$\approx$ 0–17	DC
Information redundancy based			
Single parity	1–6	$\approx$ 0–10	D
SEC-DED	7–32	10–129	DC
SNC-DND	13–75	–	DC
DEC-TED	13–69	22–200	DC

in register files (RFs); experimental results for *SEC* applied to a *64*-bit *32*-word RF using *90* nm standard cell *ASIC* technology (Naseer et al. 2006) incurs a *22%* area penalty and a *129%* increase in read access time.

TMR applied to the same type of registers incurs a *204%* area penalty but increases the read access time by only *17%*. Therefore, for sensitive *ASIC* applications that demand low-latency, *TMR* is more suitable.

3.8 Conclusion

The use of fault avoidance techniques does not guarantee complete removal of faults, having many drawbacks in terms of cost, speed of operation and chip area. System testing and verification techniques can never be exhaustive enough to remove all potential faults and their causes.

Structural redundancy techniques, such as DWC for single error detection and TMR for single error correction, are very popular. However, both techniques entail high area and power overheads and may not be suitable in embedded applications where power consumption is an important issue.

CED techniques based on EDC, such as parity coding, involve lower area overheads than structural redundancy techniques, but have limited detection abilities and cannot correct errors or efficiently detect multi-bit errors. ECC and physical interleaving incur large area overheads for multi-bit errors. Identifying the time interval for scrubbing can be tricky.

In terms of time redundancy, the aim of a cost-effective design should be finding a function C that provides a good trade-off between high coverage and low complexity. If the hardware required to implement the coding and decoding functions is similar to that of implementing $f(x)$ then structural redundancy techniques are more effective. Time redundancy cannot be used in every application due to the additional time required. For instance, certain long-life critical systems used in space applications can tolerate additional time much easier than additional space or power requirements whereas real-time safety critical systems used in avionics cannot afford any additional performance penalty.

Apart from time, extra hardware in the form of shifters, registers, comparators and extra bits are needed in ALUs. Moreover, fault coverage is not provided for shifters, rotators and comparators unless they are implemented with self-checking capabilities. However, if time is available TR techniques do offer an opportunity to minimize the additional hardware required as compared to structural redundancy.

When it comes to implementing FT, the selection of particular types of redundancy greatly depends upon the application. Therefore, to select a specific set of redundancy techniques for implementation, we should examine (a) the different requirements of the specific application and (b) the techniques that are more suitable for such requirements. Likewise, not only the type of redundancy technique is important, but where and at which level it's applied; for instance, applying TMR at the gate, register or circuit level would have a different fault coverage, time, structural and power consumption trade-off.

References

Arimoto K, Matsuda Y, Furutani K, Tsukude M, Ooishi T, Mashiko K, Fujishima K (1990) A speed-enhanced DRAM array architecture with embedded ECC. IEEE J Solid-State Circuits 25:11–17

Asakura M, Matsuda Y, Hidaka H, Tanaka Y, Fujishima K (1990) An experimental 1-Mbit cache DRAM with ECC

Avizienis A (1971) Faulty-tolerant computing: an overview. Computer 4:5–8

Avizienis A (1976) Fault-tolerant systems. IEEE Trans Comput 25:1304–1312

Avizienis A, Kelly JP (1984) "Fault Tolerance by Design Diversity: Concepts and Experiments," IEEE Computer, vol 17, no 8, August, pp 67–80

Baeg S, Wen S, Wong R (2009) SRAM interleaving distance selection with a soft error failure model. IEEE Trans Nucl Sci 56:2111–2118

Baumann RC (2001) Soft errors in advanced semiconductor devices-part I: the three radiation sources. IEEE Trans Device Mater Reliab 1:17–22

Baumann R, Hossain T, Murata S, Kitagawa H (1995) Boron compounds as a dominant source of alpha particles in semiconductor devices. In: 33rd annual proceedings reliability physics symposium, 1995. IEEE International, pp 297–302

Bentoutou Y, Djaifri M (2008) Observations of single-event upsets and multiple-bit upsets in random access memories on-board the Algerian satellite. In: Nuclear science symposium conference record, 2008. NSS'08. IEEE, pp 2568–2570

Boatella C, Hubert G, Ecoffet R, Duzellier S (2009) ICARE on-board SAC-C: more than 8 years of SEU & MCU, analysis and prediction. IEEE, pp 369–374

Bogliolo A, Favalli M, Damiani M (2000) Enabling testability of fault-tolerant circuits by means of IDDQ checkable voters. IEEE Trans Very Large Scale Integr Syst 8:415–419

Bose RC, Ray-Chaudhuri DK (1960) On a class of error correcting binary group codes. Inf Control 3:68–79

Buckle R, Highleyman WH (2003) The new nonstop advanced architecture: a massive jump in processor reliability. The Connection 24

Carter WC, Bouricius WG (1971) A survey of fault tolerant computer architecture and its evaluation. Computer 4:9–16

Carter WC, Schneider PR (1968) Design of dynamically checked computers. In: IFIP congress, no 2, pp 878–883

Cazeaux JM, Rossi D, Metra C (2004) New high speed CMOS self-checking voter. In: 10th IEEE proceedings of the international on-line testing symposium, IOLTS'04. IEEE Computer Society, Washington, DC, USA, p 58

Chen CL, Hsiao MY (1984) Error-correcting codes for semiconductor memory applications: a state-of-the-art review. IBM J Res Dev 28:124–134

Constantinescu C, Parulkar I, Harper R, Michalak S (2008) Silent Data Corruption - Myth or reality?, in: IEEE International Conference on Dependable Systems and Networks With FTCS and DCC, 2008. DSN 2008. Presented at the IEEE International Conference on Dependable Systems and Networks With FTCS and DCC, 2008. DSN 2008, pp 108–109. https://doi.org/10.1109/DSN.2008.4630077

Hamming RW (1950) Error correction and error detection coding. Bell Syst Tech J XXIX

Hana HH, Johnson BW (1986) Concurrent error detection in VLSI circuits using time redundancy. In: Proceedings of IEEE SoutheastCon 1986 regional conference, pp 208–212

Haraszti TP (2000) CMOS memory circuits. Springer. ISBN: 978-0-7923-7950-8

Hentschke R, Marques F, Lima F, Carro L, Susin A, Reis R (2002) Analyzing area and performance penalty of protecting different digital modules with hamming code and triple modular redundancy. In: SBCCI'02. IEEE Computer Society, USA, pp 95–99

Hsiao MY (1970) A class of optimal minimum odd-weight-column SEC-DED codes. IBM J Res Dev 14:395–401

Hughes HL, Benedetto JM (2003) Radiation effects and hardening of MOS technology: devices and circuits. IEEE Trans Nucl Sci 50:500–521

Johnson BW (1989) The design and analysis of fault tolerant digital systems. Addison-Wesley

Johnson BW, Aylor JH, Hana HH (1988) Efficient use of time and hardware redundancy for concurrent error detection in a 32-bit VLSI adder. IEEE J Solid-State Circuits 23:208–215

Kadayif I, Sen H, Koyuncu S (2010) Modeling soft errors for data caches and alleviating their effects on data reliability. Microprocess. Microsyst. 34, pp 200–214. https://doi.org/10.1016/j.micpro.2010.04.003

Kato M, Watanabe K, Okabe T (1989) Radiation effects on ion-implanted silicon-dioxide films. IEEE Trans Nucl Sci 36:2199–2204

Kim J, Hardavellas N, Mai K, Falsafi B, Hoe J (2007) Multi-bit error tolerant caches using two-dimensional error coding. In: 40th annual IEEE/ACM international symposium on microarchitecture, MICRO 40. IEEE Computer Society, pp 197–209

Lala JH, Harper RE (1994) Architectural principles for safety-critical real-time applications. Proc IEEE 82:25–40

Latchoumy P, Sheik P, Khader A (2011) Survey on fault tolerance in grid computing

Li J, Swartzlander E (1992) Concurrent error detection in ALUs by recomputing with rotated operands. In: 1992 IEEE international workshop on defect and fault tolerance in VLSI systems, 1992. Proceedings, pp 109–116

Lin S, Costello DJ (1983) Error control coding: fundamentals and applications. Prentice Hall
May T (1979) Soft errors in VLSI: present and future. IEEE Trans Compon Hybrids Manuf Technol 2:377–387
McEvoy D (1981) The architecture of Tandem's nonstop system. In: Proceedings of the ACM'81 conference, ACM'81. ACM, New York, NY, USA, p 245
Metra C, Favalli M, Ricco B (1997) Compact and low power on-line self-testing voting scheme. In: 1997 IEEE international symposium on defect and fault tolerance in VLSI systems, 1997. Proceedings, pp 137–145
Moon TK (2005) Error Correction Coding: Mathematical Methods and Algorithms. John Wiley & Sons.
Mrstik BJ, Hughes HL, McMarr PJ, Lawrence RK, Ma DI, Isaacson IP, Walker RA (2000) Hole and electron trapping in ion implanted thermal oxides and SIMOX. IEEE Trans Nucl Sci 47:2189–2195
Mukherjee SS, Emer J, Fossum T, Reinhardt SK (2004) Cache scrubbing in microprocessors: myth or necessity? In: PRDC'04. IEEE Computer Society, Washington, DC, USA, pp 37–42
Naseer R, Bhatti RZ, Draper J (2006) Analysis of soft error mitigation techniques for register files in IBM Cu-08 90 nm technology. In: IEEE international Midwest symposium on circuits and systems, 2006. MWSCAS'06, pp 515–519
Neuberger G, de Lima F, Carro L, Reis R (2003) A multiple bit upset tolerant SRAM memory. ACM Trans des Autom Electron Syst 8:577–590
Neuberger G, de Lima Kastensmidt FG, Reis R (2005) An automatic technique for optimizing Reed–Solomon codes to improve fault tolerance in memories. IEEE des Test Comput 22:50–58
Nishioka Y, Ohyu K, Ohji Y, Kato M, da Silva EF, Ma TP (1989) Radiation hardened micron and submicron MOSFETs containing fluorinated oxides. IEEE Trans Nucl Sci 36:2116–2123
Patel JH, Fung LY (1982) Concurrent error detection in ALU's by recomputing with shifted operands. IEEE Trans Comput 31:589–595
Paul S, Cai F, Zhang X, Bhunia S (2011) Reliability-driven ECC allocation for multiple bit error resilience in processor cache. IEEE Trans Comput 60:20–34
Pellish JA, Reed RA, Schrimpf RD, Alles ML, Varadharajaperumal M, Niu G, Sutton AK, Diestelhorst RM, Espinel G, Krithivasan R, Comeau JP, Cressler JD, Vizkelethy G, Marshall PW, Weller RA, Mendenhall MH, Montes EJ (2006) Substrate engineering concepts to mitigate charge collection in deep trench isolation technologies. IEEE Trans Nucl Sci 53:3298–3305
Puchner H, Kapre R, Sharifzadeh S, Majjiga J, Chao R, Radaelli D, Wong S (2006) Elimination of single event latchup in 90 nm SRAM technologies. In: 44th annual reliability physics symposium proceedings, 2006, pp 721–722
Ravishankar K, Iyer ZK (2003) Hardware and software error detection
Reed IS, Solomon G (1960) Polynomial codes over certain finite fields. J Soc Ind Appl Math 8:300–304
Reed RA, Carts MA, Marshall PW, Marshall CJ, Musseau O, McNulty PJ, Roth DR, Buchner S, Melinger J, Corbiere T (1997) Heavy ion and proton-induced single event multiple upset. IEEE Trans Nucl Sci 44:2224–2229
Reviriego P, Maestro JA, Cervantes C (2007) Reliability analysis of memories suffering multiple bit upsets. IEEE Trans Device Mater Reliab 7:592–601
Reviriego P, Maestro JA, Baeg S, Wen S, Wong R (2010) Protection of memories suffering MCUs through the selection of the optimal interleaving distance. IEEE Trans Nucl Sci 57:2124–2128
Reynolds DA, Metze G (1978) Fault detection capabilities of alternating logic. IEEE Trans Comput 27:1093–1098
Roche P, Gasiot G (2005) Impacts of front-end and middle-end process modifications on terrestrial soft error rate. IEEE Trans Device Mater Reliab 5:382–396
Saleh AM, Serrano JJ, Patel JH (1990) Reliability of scrubbing recovery-techniques for memory systems. IEEE Trans Reliab 39:114–122
Schagaev I (1989) Yet another approach to classification of redundancy. In: Proceedings of FTSD, Prague, Czechoslovakia, pp 485–490

Schagaev I, Buhanova G (2001) Comparative study of fault tolerant RAM structures. In: IEEE DSN01, Goteborg
Schagaev I, Sogomonian E (1988) Hardware and software fault tolerance of computer systems. Avtomat Telemekhan 3–39
Schagaev IV, Bernstein AV, Tomfield YL (1992) Storage unit with high reliability characteristics. I. Avtomat Telemekh 145–152
Schagaev IV, Bernstein AV, Tomfield YL (1993) RAM of high reliability properties. II. Avtomat Telemekh 169–179
Schagaev I et al (2001) Redundancy classification and its applications for fault tolerant computer design. In: IEEE TESADI-01
Schwank JR et al (2003) Charge collection in SOI capacitors and circuits and its effect on SEU hardness. IEEE Trans Nucl Sci 2937–2947
Shannon CE (1948) A mathematical theory of communication. Bell Syst Tech J 27:379–423, 623–656
Shedletsky JJ (1978) Error correction by alternate-data retry. IEEE Trans Comput 27:106–112
Sherman L (2003). Stratus continuous processing technology—the smarter approach to uptime. Whitepaper. Technical report. Stratus Technologies
Sklaroff JR (1976) Redundancy management technique for space shuttle computers. IBM J Res Dev 20:20–28
Slayman CW (2005) Cache and memory error detection, correction, and reduction techniques for terrestrial servers and workstations. Device and Materials Reliability, IEEE Transactions on 5, pp 397–404. https://doi.org/10.1109/TDMR.2005.856487
Slegel TJ, Averill RM, Check MA, Giamei BC, Krumm BW, Krygowski CA, Li WH, Liptay JS, MacDougall JD, McPherson TJ, Navarro JA, Schwarz EM, Shum K, Webb CF (1999) IBM's S/390 G5 microprocessor design. IEEE Micro 19:12–23
Stroud CE (2002) A Designer's to Built-in Self-Test. Springer. https://link.springer.com/book/10.1007/b117480
Von Neumann J (1956) Probabilistic logics and the synthesis of reliable organisms from unreliable components. In: Automata studies. Princeton University Press, pp 43–98
Weaver C, Emer J, Mukherjee SS, Reinhardt SK (2004) Techniques to reduce the soft error rate of a high-performance microprocessor. In: 31st annual international symposium on computer architecture, proceedings, pp 264–275
Williams TW, Kapur R, Mercer MR, Dennard RH, Maly W (1996) Iddq testing for high performance CMOS-the next ten years. In: European design and test conference, 1996. ED TC 96
Wood A (1999) Data integrity concepts, features, and technology. White paper, Tandem Division, Compaq Computer Corporation
Woodard SE, Metze G (1978) Self-checking alternating logic: sequential circuit design. In: Proceedings of the 5th annual symposium on computer architecture, ISCA'78. ACM, New York, NY, USA, pp 114–122
Yeh YC (1996) Triple-triple redundant 777 primary flight computer. In: 1996 IEEE aerospace applications conference. Proceedings, Aspen, CO, USA, vol 1, pp 293–307. https://doi.org/10.1109/AERO.1996.495891

Chapter 4
FT Models of Fault Tolerance

Abstract Models of faults and methods of toleration of faults were presented in some detail and in concert. Classification of both enables to rigorously design system with adequate level of fault toleration. Fault appearance, manifestation and toleration should be considered and analysed as processes required to achieve high level of maintainability and system resilience. Some fault hierarchy is discussed arguing that toleration scheme must be designed to tolerate "the heaviest".

Keywords Model of fault · Fault tolerance · Resilience · Redundancy types · Generalised algorithm of fault tolerance · Reliability model

4.1 Models

We define *M* as the known model of a system that performs a given function *F*. Let's imagine a new feature of extreme reliability in that model. In order to express the existence of this new feature, the predicates *P* and *Q* are introduced to determine the state of the model. *P* and *Q* also defined the direction of the time arrow (see Fig. 4.1).

The model of this kind we were using in our book (Schagaev and Gutknecht 2024) introducing possible change of inner state of the model by external, or internal impacts. Readers who are interested in this aspect might find more material about this kind of model there.

To analyse methods for achieving a required level of reliability with performance and power consumption constraints, we offer a combination of the following three models:

1. The model of the system M_s
2. The model of the faults M_{fault} that a *RT FT* system will be exposed to
3. The model of fault tolerance M_{FT} or the new structure that implements *FT*

As Fig. 4.1 shows, M_s, M_{fault} and M_{FT} are mutually dependent models. Notice that in this approach development and manufacturing costs of a solution are not considered.

© The Author(s), under exclusive license to Springer Nature Switzerland AG 2026
I. Schagaev, *Resilience of Computer Systems*,
https://doi.org/10.1007/978-3-032-00938-8_4

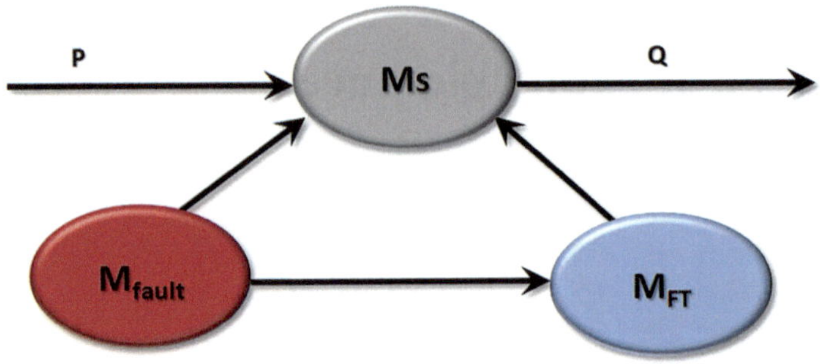

Fig. 4.1 New feature of an *FT* system: reliability

Table 4.1 Typical example of HW faults

Type of fault	Description	Impact
Byzantine	The behaviour of a component that gives conflicting values to other components	The entire system is affected
Subsystem fault	A temporary or permanent incorrect behaviour of a subsystem	The entire system is affected
Open fault	Resistance on either a line or a block due to a bad connection	The value associated to the line or the block is modified
Bridging fault	Signals S_1 and S_2 are connected unintentionally	The value associated to the line or the block is modified to a different value
Stuck-at fault	The result value is fixed to 0 or 1	The result value is stuck to 0 or 1
Bit-flip fault	A state switch from 0 to 1, or vice versa, when it should not	The result changes its original value

M_{fault} is a description of all faults that a system must tolerate. In binary logic, a typical permanent fault can manifest as "stuck at zero" or "stuck at one".

Table 4.1 shows typical examples of HW faults. Hidden faults, also called Latent faults, are behavioural faults that exist in the hardware over a long period of time, e.g.: Byzantine faults[1] and fail-stop[2] faults. Both types complicate the design of *FT*; all described faults should be tolerated within a limited and specified period of time. This period actually determines the availability of the system. Fault types differ by their impact, as well as the way they are handled.

[1] Byzantine faults occur when a faulty system continues to operate, producing incorrect results sometimes giving the impression that they are working correctly. Dealing with this type of fault is difficult.

[2] Fail-stop (also known as fail-silent) faults take place when a faulty unit stops functioning producing no bad output. It either produces not output or produces correct results that clearly indicate that the unit has failed.

Thus, the fault model has its own hierarchy, including single-bit, element, behavioural and subsystem faults. One has to accept that the fault type is varying and some action hierarchy to tolerate them is also required. A detailed fault model is further developed and discussed in further sections.

Fault encapsulation approaches can help to handle faults: due to deliberate design solutions, it is possible to ensure that severe faults in the system do not escalate and remain simpler to handle; therefore, making the fault handling practically possible to implement.

RT FT system applications assume long operational life; however, fault-handling schemes are needed much more often towards the end of the device life cycle. The appropriate techniques for tolerating faults of various types are also presented in Table 4.2.

As discussed in previous chapters, to tolerate transient faults, one can use time redundancy schematic support imbedded in hardware (e.g. instruction re-execution). System software support is also needed, as the hardware cannot cover all possible faults.

It is obvious that faults, occurring at the bit level (stuck zero, stuck one and similar) should be efficiently handled *ASAP* (as soon as possible) and *ALAP* (as local as possible), i.e. at the same or nearest level. The term "*level*" in our case means the level in the hardware hierarchy on which the fault should be handled.

In other words, when a "*stuck-at zero*" permanent fault has occurred in the register file (RF) with no corrective schemes available, the whole RF has to be replaced, if no other possible reconfigurations were predefined.

In turn, when only one RF is integrated in the chip and no other reconfigurations are defined, then the whole chip has to be replaced, etc.

Pursuing these two principles allows limiting the fault spreading and its impact to a higher level either in the chip or the system as a whole.

Example: To tolerate bit-flip faults, hardware and system software information redundancies might be used, as well as hardware structural support. In this sense, parity checking in registers, supported and implemented concurrently by hardware, is described as *HW(δI)*. *HW(δS)* and *HW(δT)* are needed as supportive redundancies, *HW(δS)* describing the additional parity line and comparison logic, and *HW(δT)* the additional time needed to update the parity line and executing the comparison.

However, the main type of redundancy used in this approach is information.

An exact characterization of the distribution of faults for computer systems is extremely difficult due to the number of different factors that determine faults, such temperature, vibration, radiation exposure, etc. Besides, discriminating between transient and permanent faults is difficult.

The transient-permanent fault ratio varies from 10 to more than 1000 depending on the technology, manufacturing scale, operating conditions, etc. In the case of memory, a typical value of hard error rates is in the order of 10–100, whereas for soft errors it can vary between 1000 and more than 5000 for the same period of testing or operation.

The upper bound belongs to aerospace and aviation, principally due to faults induced by alpha particles.

Table 4.2 Classification of fault by origin

Level of abstraction	Structural faults	Transistor level (component faults)	Stuck-open or stuck-off
			Stuck-short or stuck-on
		Gate level (interconnect faults)	Stuck-at (Armstrong 1966; Galey et al. 1961): s-a-0, s-a-1
			Timing delay (Smith 1985): path-delay (PDF) and gate-delay (GDF) faults
			Bridging (Mei 1974)
	Functional faults	Pattern sensitive faults (PSF) (Hayes 1975)	Passive PSF
			Active PSF
			Static PSF
		Coupling faults (CF)	
Phase of creation of occurrence (Landwehr et al. 1994)	Development faults	e.g. specification faults, implementation and manufacturing faults	
	Operational faults	e.g. aging faults e.g. alpha particle hits	
System boundaries (Avizienis et al. 1984)	Internal	e.g. design and implementation faults	
	External	e.g. radiation, temperature changes, power surges from external power supply	
Phenomenological cause	Natural (Jennings 1990)		
	Human-made (Hugue and Purtilo 2002)		
Capability/objective/ intent (Brocklehurst et al. 1994)	Malicious	Deliberate	
	Non-malicious	Accidental	
		Incompetence	
Nature (Avizienis et al. 1984)	Hardware		
	Software		
Cause	Specification mistakes		
	Defects	Implementation mistakes	e.g. Pentium FDIV bug (Coe et al. 1995; Price 1995)
		Manufacturing defects	Global defects or Gross area defects (Koren and Koren 1998)
			Spot defects (Koren and Singh 1990)
	Operating environment External disturbances	Thermal stress	
		Heat	
		Electro-migration EM	
		Voltage drop	

(continued)

Table 4.2 (continued)

		Noise	Electrical overstress	Hot carrier injection HCI (DAHC, CHE, SHE, SGHE)
				Negative Bias temperature instability (NBTI)
				Latchup
			Induced charging	
			Oxide Breakdown	
			Radiation	See Table 4.1
			EMP	

Figures 3.2 and 4.1 are transformed into Fig. 4.2, which presents various faults in the system and various possible solutions. M_{fault} illustrates the fact that the fault types are not separated.

For example, Byzantine faults of the system might be "*stuck at zero*" faults of the hardware that were spread throughout the system. The latency of faults thus becomes crucial in determining the reliability of the system. Consequently, different faults require different actions and mechanisms to tolerate them.

The system model of Fig. 4.2 has overlapped *SSW* and *HW* ellipses to represent the duality of the system: hardware and system software. Both of them must be involved in the implementation of fault tolerance and real-time features.

The overlapped *HW* and *SSW* ellipses indicate that HW and SSW functions might be applied to tolerate specific types of hardware faults. Other fault types might also

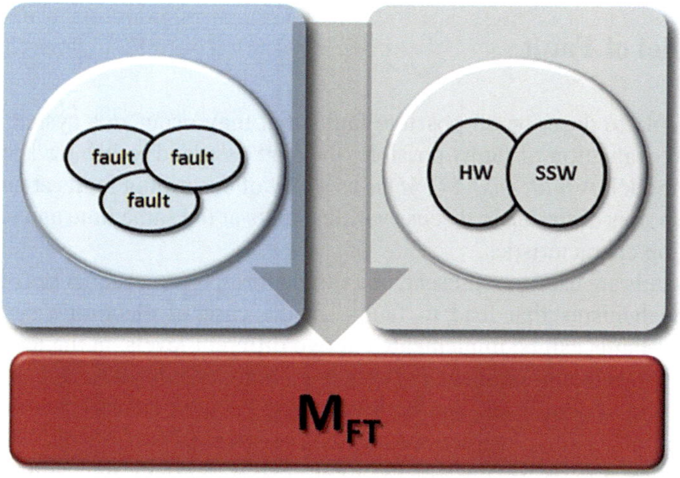

Fig. 4.2 Fault tolerance: a model of a computer system

be tolerated by HW or SSW only. M_{ft} is "a conceptual deliverer" of reliability for the *RT FT* system. It has to be effective during the whole operational lifetime of the computer itself.

As hardware degrades over time, the fault tolerance mechanisms are more likely to be used towards the end of the life cycle. FT systems are designed with the assumption that new types of faults do not appear during the operational life-time of the system, i.e. the system must be designed to be fault-tolerant for the set of faults and their types known at design time. All these solutions require careful analysis due to their impact on the system reliability.

In contrast to the usual assumption in reliability modelling, one has to assume that a fault might exist in the system over an arbitrary long period of time (latent fault) and its detection and elimination is not possible "at once". Consequently, we accept that FT is a process, and discuss it in the following sections. Using Dijkstra's approach (Dijkstra 1965) of defining a function as a process described by its algorithm, we consider FT as a function that is also described and implemented by an algorithm.

There are several options to achieve fault tolerance assuming the use of *HW* and *SSW* by using various types of redundancy mentioned above. However, the use of certain redundancy types might cause system performance degradation, which is especially true for software measures (Kulkarni et al. 1987; Oh et al. 2002). Thus, the analysis of performance/reliability dependency, impact and degradation should be taken into account.

The introduced system redundancy might be used in a way to tolerate only certain fault types, thus degrading fault coverage, keeping performance at acceptable levels. Software-based redundancy might preserve the same type of fault coverage but with more time redundancy—delays (recovery time degrades, availability degrades), but the fault coverage might not, thus the system degrades in terms of reliability.

4.2 Model of Fault

It is unfeasible to describe all possible faults that may occur in a system. In order to make the evaluation of faults possible, they are assumed to behave according to some fault model. A *fault model* (*FM*) is as a way of describing faults at once (Dunn 1991). Often, it is desirable to discuss various faults at the same time and to describe their common characteristics.

Fault models are used to represent, in a simple form, the consequences of complex physical mechanisms that lead to faults. In the case of electronic systems, the modelling of faults can be implemented at two different levels: at the level of hardware components that implement a system (e.g. memory subsystems, register banks, *ALU*) or at the system level, that is directly related with the information that the system manipulates (e.g. instructions and data program).

The simplest *FM* is to consider the logic gate as a single unit with a constant failure rate, instead of considering different failure rates for the individual transistors that form the unit. As in Table 4.2, analysing the physics of faults to the atomic and molecular level would provide a clear understanding of the failure mechanisms. Such understanding is very helpful in the development of fault models. Advancing from (Avizienis et al. 1984), we extend the classification of faults depending on the ways how fault are originated or manifested.

4.3 Classification of Faults by Origin

Faults can be classified differently depending on attributes related to their origin, including their cause, the level at which they take place, the phase of creation, nature, system boundaries, phenomenological causes and intention. Table 4.2 shows a number of faults classified by their origin attributes.

4.3.1 Level Response

Hardware defects can be the source of *physical faults*. *Logical faults* can be used to model the manifestation of physical faults on the behaviour of a system. Logical faults represent physical faults in order to simplify the logic function of the circuit and fault analysis process. They can be subdivided into *structural faults*, which are related to structural models and modify the interconnection among components, and *functional faults*, which are related to functional models and change the functions of components and circuits.

Component faults are a type of structural faults, which can be applied at the *transistor level*. Some of these are:

1. stuck-open or stuck-off: a transistor is always off and not controllable by gate input
2. stuck-short or stuck on: a transistor is always on and not controllable by gate input.

Another type of structural faults, *interconnect faults*, can be applied at the *gate level*. Among these:

1. stuck-at faults (SAF): single or multiple lines have a constant value of 0 (s-a-0 faults) or 1 (s-a-1 faults) regardless of the value of the other signals in the circuit (Armstrong 1966; Galey et al. 1961)
2. timing/delay faults: certain defects due to manufacturing or external reasons do not change the logic function of components, but can cause timing violations; faults due to propagation delays along a path (path-delay faults, PDF) or gate (transition or gate delay fault, GDF) (Smith 1985), exceeding the limits required for correct operation
3. bridging faults (BFs): two or more distinct lines are shorted (Mei 1974) usually due to particles or shorted metal line. Depending on the value, the bridging could

be AND bridging (also referred to as 0-dominant), OR bridging (also referred to as 1-dominant) or Indeterminate Bridging. It is obvious that the probability of BF's occurring increases with (1) shorter distances between metal lines due to the use of shrinking technology and (2) the use of long parallel lines (Tehranipoor et al. 2012).

Different types of functional faults that can be applied at the functional level are:

1. pattern sensitive faults (PSFs): where a fault signal depends on the signal values of nearby components (Hayes 1975); typical in DRAM, there are three types of PSFs due to changes in the neighbourhood pattern:
   (a) Passive PSF: the value of a cell remains
   (b) Active PSF: the value of a cell changes
   (c) Static PSF: the value of a cell is being forced to a particular state (0/1)
2. coupling faults (CF): A subset of SPF, representing a specific pattern sensitivity between two memory cells (Nair et al. 1978); Two memory cells C_j and C_j are coupled if a transition from X to Y in one cell, say C_i, changes the state of the other cell, given that:

$$X \in \{0, 1\} \quad \text{and} \quad X = \overline{A}$$

3. Idempotent coupling faults: a transition $0 \rightarrow 1, \text{ or } 1 \rightarrow 0$ in Ci forces the contents of C_j to a specific value $X \in \{0, 1\}$
4. Inversion coupling faults: a transition $0 \rightarrow 1, \text{ or } 1 \rightarrow 0$ in Ci forces an inversion $0 \rightarrow 1, \text{ or } 1 \rightarrow 0$ of C_j.

4.3.2 *Cause of Faults*

4.3.2.1 Specification Mistakes

Specification mistakes, which take place during the planning and design phases, can be the source of faults (*specification faults*), including incorrect timing, power and environmental requirements. The effect of certain specification faults may be corrected via fault masking.

4.3.2.2 Defects

A hardware defect in electronics is the unintended difference between the implementation and the intended design. *Implementation mistakes*, such as the Pentium FDIV bug (Coe et al. 1995; Price 1995), are a type of defects. Conversely, imperfections in the fabrication process of state-of-the-art VLSI technologies result in *manufacturing defects*, whose severity increases proportionally with the size and density of the chip.

Manufacturing defects are largely dependent on the specific technology and layout, and include processing and material defects such: dust particles on the chip, conducting layer defects (shorts and opens), oxide defects, scratches and gate oxide pinholes, defects caused by either extra or missing material (Koren and Koren 1998). Manufacturing defects can be classified as *global defects* (or *gross area defects*), affecting large areas of a wafer and so can be easily detected during manufacturing, or as *spot defects*, which are random, affecting areas comparable to the single device size, and therefore more difficult to be detected (Koren and Singh 1990).

4.3.2.3 Operating Environment

Correct functioning of digital systems is based on the assumption that electrical and timing transistor parameters will remain bounded to certain margins (usually ± 15%). This margin tolerance specified at initial manufacturing can be violated during operating time due to shifts induced by external disturbances.

These mechanisms can produce systematic degradation overtime or abrupt failures of basic components. Transistors can be degraded due to electrical overstress and radiation whereas oxide-breakdown, electrostatic discharge and ionizing radiation are usually the cause of abrupt failures.

Hot carrier injection (*HCI*) has been one of the most common electrical overstress aging mechanism, adversely affecting both *nMOS* and *pMOS* transistors. It occurs when a charge carrier, an electron or a hole, gain enough kinetic energy to break an interface state.

Different mechanisms can be responsible for *HCI,* including s*ubstrate hot electrons* (*SHE*) (Ning and Yu 1974), c*hannel hot electrons* (*CHE*) (Cottrell et al. 1979), *drain avalanche hot carriers* (*DAHC*) (Takeda et al. 1983) and *secondarily generated hot electrons* (*SGHE*) (Matsunaga et al. 1980).

Negative Bias temperature instability (*NBTI*) (Schroder and Babcock 2003) is also a critical reliability concern for p*MOS* transistors (not so much for n*MOS)* and has been a persistent issue for generations below 130 nm (Schroder and Babcock 2003; Alam et al. 2007). Interface traps are generated during negative bias conditions ($V_{gs} = -V_{dd}$). Higher temperatures seem worsen *NBTI*, producing larger voltage, which if maintained over long periods (*NBTI* exhibits logarithmic dependence on time), may significantly increase delays (Kumar et al. 2006; Kaczer et al. 2005).

Another example of electrical-overstress mechanism is the *latchup* can also be triggered electrically (Gregory and Shafer 1973).

In turn, *non-ionizing radiation* can be the cause of *DDD*, while *TID* effects can be induced by ionizing radiation. Other degradation mechanisms can affect interconnection logic, e.g. *electro-migration* (*EM*).

In contrast with the previous long-term degradation mechanisms, the effect of noise can produce abrupt failures. Examples of these are faults induced by the effects of noise including *oxide breakdown*, *electrostatic discharge* (*ESD*) and *ionizing radiation*.

Oxide Breakdown is the destruction of an oxide layer of a semiconductor device, e.g. *time dependent dielectric breakdown (TDDB)*, *early-life dielectric breakdown (ELDB)*, and *EOS/ESD-induced dielectric breakdown.*

4.3.3 Phase of Formation and Occurrence of Faults

Faults that take place during the manufacturing phase are *development faults* in contrast with *operational faults* that take place during the service delivery of the operation phase (Landwehr et al. 1994), e.g. faults due to radiation.

4.3.4 Nature/Dimension

According to their nature, faults can be categorized as *hardware* (such as combinational and sequential logic defects due to aging, radiation, etc.) or *software* (branch errors, missing instructions and pointers, etc.). The scope of this thesis focuses exclusively on hardware faults and their effects.

4.3.5 System Boundaries

With respect to the system boundaries, faults can also be classified as internal (originate inside the system boundary) or external (originate outside the system boundary).

Internal faults are those that arise from within a system, often due to design flaws. These are usually repeatable for a given set of inputs in the system. In addition, they can also be the result of implementation faults, which if random, are difficult to repeat.

External faults are those that originate from outside the system, propagating into the system. These are normally the result of interference caused by the physical environment, including *environmental faults* (e.g. radiation, temperature changes), accidental damage from an external system (e.g. power surges from an external power supply), etc.

4.3.6 Phenomenological Cause

The key components of embedded systems have an inherent susceptibility to Natural (Jennings 1990) and human-made (Hugue and Purtilo 2002) faults. Natural faults are generally random in nature and are caused by natural phenomena, without human participation.

These are normally a consequence of environmental overstress. Human-made faults are the result of human action, including design and interaction faults (operational misuses), and are usually due to mistakes in the design, implementation, or use of systems.

4.3.7 Capability/Objective/Intent

Following the previous classification, human-made faults can either be deliberately harmful (*malicious faults*) or can be triggered without purpose and awareness (*non-malicious faults*) (Brocklehurst et al. 1994). *Accidental faults* are due to mistakes and bad decisions as long as they are not made deliberately; these include interaction, design and implementation faults.

It is obvious that all natural faults have no intention and, therefore, are accidental. *Incompetence faults* are faults due to mistakes or bad decisions that were the result of a lack of professional competence.

4.4 Conclusion

In this chapter, we have discussed models of system, models of fault in combination with models of fault toleration.

The latter was not only about static analysis, as we consider fault toleration as a process prepared during the design and executed during the whole period of operational life of a system.

The distinction of this chapter from other works about fault models lies in detailed presentation of a fault for each chosen example, and generalized scheme of fault tolerance—in other words, a detailed fault model analyzed and handled together with generalized methods and models of fault toleration.

We have analysed specification of fault caused by design mistake, hardware defect, impact of operating environment. We define faults as internal, originated inside the system boundaries and the ones that go beyond them.

Causes of random or deliberate fault were also differentiated.

We have extended the model of faults defined by (Laprie et al. 2004) and provided a classification suggesting methods for recognition and reaction. Manifestation, detectability, diagnosability and recoverability are discussed as one consistent flow proposing adequate solutions for diagnosis and recovery.

Acknowledgements As before mentioned—this chapter is a summary of our discussion with Dr Victor Castano, known as well as new models of fault tolerance, made over several years. His involvement is highly appreciated.

References

Alam MA, Kufluoglua H, Varghese D, Mahapatra S (2007) A comprehensive model for PMOS NBTI degradation: recent progress. Microelectronics Reliab 47:853–862

Armstrong DB (1966) On finding a nearly minimal set of fault detection tests for combinational logic nets. IEEE Trans Electron Comput EC-15:66–73

Avizienis A, Kelly JP (1984) "Fault Tolerance by Design Diversity: Concepts and Experiments," IEEE Computer, vol 17, no 8, August, pp 67–80

Brocklehurst S, Littlewood B, Olovsson T, Jonsson E (1994) On measurement of operational security. IEEE Aerosp Electron Syst Mag 9:7–16

Coe T, Mathisen T, Moler C, Pratt V (1995) Computational aspects of the Pentium affair. IEEE Comput Sci Eng 2:18–30

Cottrell PE, Troutman RR, Ning TH (1979) Hot-electron emission in N-channel IGFET's. IEEE Trans Electron Dev 26:520–533

Dijkstra EW (1965) Solution of a problem in concurrent programming control. Commun ACM 8:569

Dunn M (1991) Designer fault models for VLSI. In: IEE colloquium on design for testability, pp 4/1–4/5

Galey JM, Norby RE, Roth JP (1961) Techniques for the diagnosis of switching circuit failures. In: Proceedings of the second annual symposium on switching circuit theory and logical design, 1961 (SWCT 1961), pp 152–160

Gregory BL, Shafer BD (1973) Latch-up in CMOS integrated circuits. IEEE Trans Nucl Sci 20:293–299

Hayes JP (1975) Detection of pattern-sensitive faults in random-access memories. IEEE Trans Comput C-24:150–157

Hugue MM, Purtilo J (2002) Guerrilla tactics: motivating design patterns for high-dependability applications. In: 27th annual NASA Goddard/IEEE software engineering workshop, 2002. Proceedings, pp 33–39

Jennings BF (1990) Fault detection in microprocessor based systems. In: IEE colloquium on fault tolerant techniques, pp 7/1–718

Kaczer B, Arkhipov V, Degraeve R, Collaert N, Groeseneken G, MGoodwin M (2005) Disorder-controlled-kinetics model for negative bias temperature instability and its experimental verification

Koren I, Koren Z (1998) Defect tolerance in VLSI circuits: techniques and yield analysis. Proc IEEE 86:1819–1838

Koren I, Singh AD (1990) Fault tolerance in VLSI circuits. Computer 23:73–83

Kulkarni GV, Nicola FV, Trivedi SK (1987) Effects of checkpointing and queueing on program performance. Duke University, Durham, NC, USA

Kumar SV, Kim CH, Sapatnekar SS (2006) An analytical model for negative bias temperature instability. In: 2006 IEEE/ACM international conference on computer aided design, San Jose, CA, USA, pp 493–496. https://doi.org/10.1109/ICCAD.2006.320163

Landwehr CE, Bull AR, McDermott JP, Choi WS (1994) A taxonomy of computer program security flaws. ACM Comput Surv 26:211–254

Laprie J-C, Avizienis A, Randell B, Landwehr C (2004) Basic concepts and taxonomy of dependable and secure computing. IEEE Trans Dependable Secure Comput 1:11–33

Matsunaga J, Momose H, Iizuka H, Kohyama S (1980) Characterization of two step impact ionization and its influence in NMOS and PMOS VLSI's. In: 1980 international electron devices meeting, pp 736–739

Mei KCY (1974) Bridging and stuck-at faults. IEEE Trans Comput C-23:720–727

Nair R, Thatte SM, Abraham JA (1978) Efficient Algorithms for Testing Semiconductor Random-Access Memories. IEEE Transactions on Computers C-27:572–576

Ning TH, Yu HN (1974) Optically induced injection of hot electrons into SiO_2. J Appl Phys 45:5373–5378

Oh N, Shirvani PP, McCluskey EJ (2002) Control-flow checking by software signatures. IEEE Trans Reliab 51:111–122

Price D (1995) Pentium FDIV flaw-lessons learned. IEEE Micro 15:86–88

Schagaev I, Gutknecht J (2024) Software design for resilient computer systems. Springer. https://doi.org/10.1007/978-3-031-55139-0

Schroder DK, Babcock JA (2003) Negative bias temperature instability: road to cross in deep submicron silicon semiconductor manufacturing. J Appl Phys 94:1–18

Smith GL (1985) Models for delay faults based on paths. In: Proceedings of the international test conference, pp 342–349

Takeda E, Kume H, Nakagome Y, Makino T, Shimizu A, Asai S (1983) An As-P double diffused drain MOSFET for VLSI's. IEEE Trans Electron Dev 30:652–657

Tehranipoor M, Peng K, Chakrabarty K (2012) Delay test and small-delay defects. Test and diagnosis for small-delay defects. Springer, New York, pp 21–36

Test Method for beam accelerated soft error rate (2007)

Chapter 5
Classification of Faults by Manifestation, GAFT

Abstract Additionally to the classification of faults by appearance, and/or manifestation, this chapter presents in detail types of redundancies one can apply to tolerate a fault in the system. It is argued that redundancy of the designed system should be applied for all three—(1) performance, (2) reliability and (3) energy efficiency of the system. System redundancy therefore must be traded for the sake of all three. We declare that well-designed system must be capable to trade. We demonstrate that system fault tolerance and resilience are not longer static requirements, but on-going processes thought he whole cycle of design and operation of any system. For this, we introduce Generalized Algorithm of Fault Tolerance. Shown how model of system reliability should be redesigned, including proposed algorithm.

Keywords Type of faults · Fault manifestation · Redundancy type · Generalized algorithm of fault tolerance · Classification of fault-tolerant systems

5.1 Introduction

Apart from the origin of fault, faults can be classified based on attributes related to their manifestation, including their response, dimension, reproducibility, extent, persistence, value, detectability, etc. To summarize it, to some extent, meaningful Table 5.1 shows a number of faults classified by their manifestation attributes.

Above this, fault should be considered in terms of duration and our ability to respond it.

5.2 Response-Timeliness

Let a component C (see Fig. 5.1) receive a nonempty input sequence ($V_i \neq \text{null}$), consistent with the specification, at time T_i. For V_i, the response V_j at time T_j is correct iff:

© The Author(s), under exclusive license to Springer Nature Switzerland AG 2026

I. Schagaev, *Resilience of Computer Systems*,
https://doi.org/10.1007/978-3-032-00938-8_5

Table 5.1 Classification of fault by manifestation

Response-timeliness (Qian 2008)	Omission faults		
	Commission faults		
Hardware			
Dimension	Software		
	System		
Activation reproducibility (Laprie et al. 2004)	Solid Elusive	E.g. pattern sensitive faults (effects of temperature, delay in timing due to parasitic capacitance)	
Extent	Local global		
Persistence/duration	Permanent	Easiest to diagnose; Once the component fails, it will never work correctly again	
	Temporary	Transient intermittent	
Value	Determinate Indeterminate		
Plurality	Single		
Correlation	Multiple Independent Related		
Damage (see Table 4.2)	Soft	Transient-soft	
		Static-soft	
	Hard Pseudo-hard		
Status	Dormant		
	Active		
Prospect (Laprie 2008)	Foreseen		
	Unforeseen		
	Foreseeable		
Seriousness	Benign		
	Malicious (Meyer and Pradhan 1991)	Omissive Symmetric Transmissive Transmissive (byzantine) Asymmetric (Thambidurai and Park 1988) Strictly omissive (Azadmanesh and Kieckhafer 2000)	
Detectability (Pomeranz and Reddy 1993)	Detectable	Recoverable DRE	(Kadayif et al. 2010; Weaver et al. 2004)
	Undetectable	Operationally redundant Unrecoverable can lead to DUE and SDC (Kadayif et al. 2010; Weaver et al. 2004)	
	Partially detectable	Under certain conditions, can be detectable and irredundant. Can lead to can lead to DUE and SDC	
Diagnosability Containability Recoverability			

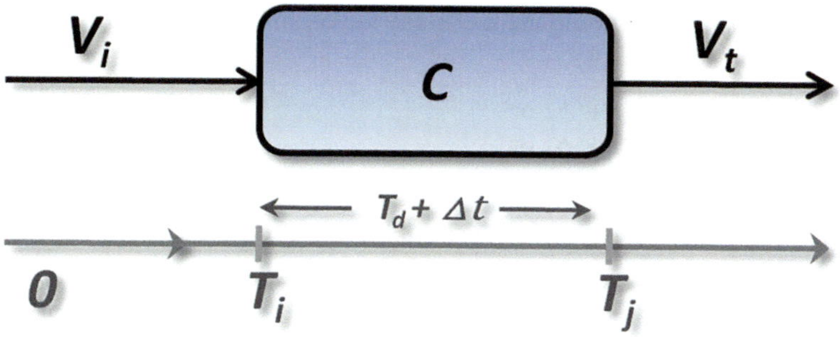

Fig. 5.1 Input-response mechanism of component *C* with single output

1. $V_j = W_j$ at time T_j, where W_j is the expected value according with the specification and
2. $T_j = T_i + T_d + \Delta t$, where T_d is the minimum delay time of the component, Δt is unpredictable delay time such that $0 \leq \Delta t \leq T_{\max}$, given that $T_{\max}$ is the maximum unpredictable delay of *C*.

Using the previous definition of *correct response* by (Qian 2008), there can be four ways with regard to timeliness and expected value, by which a response can deviate from the specification, which leads to the following classification of faults: omission, timing, timely and commission faults.

Omission faults involve the absence of actions when these should be performed. A fault that causes a component *C* not to respond to a nonempty input sequence ($V_i \neq$ null) is an *omission fault*. The potential resulting failure would be an *omission failure*, whose response would have the following properties:

1. $V_j =$ null, $T_j = T_i + T_d + \Delta t$ and
2. $V_j = W_j$, $T_j = \infty$.

A *timing fault* is a fault that causes a component *C* to respond with the expected value W_j to a nonempty input sequence ($V_i \neq$ null) either too early or too late. The corresponding failure would be a *timing failure*. Using the previous mathematical notation:

$$V_j = W_j, \text{ either } T_j < T_i + T_d \text{ or } T_j > T_i + T_d + T_{\max}$$

A *timely fault* is a fault that causes a component C to respond to a nonempty input sequence ($V_i \neq$ null), within the specified time interval, but with a wrong value. The corresponding failure would be a *timely failure*:

$$V_j \neq W_j, \ T_j = T_i + T_d + \Delta t$$

Therefore, an omission fault is also timely fault with a null value produced on time.

A commission fault of a component C is any violation from its specified behaviour, with the following properties:

1. $V_j \neq W_j, T_j = T_i + T_d + \Delta t$ or
2. $V_j \neq W_j, T_j \neq T_i + T_d + \Delta t$ or
3. $V_j = W_j, T_j \neq T_i + T_d + \Delta t$.

Consequently, a commission fault is a subset of all other three types of faults.

5.2.1 Consistency

Before classifying faults with respect to consistency, the definition of correct response was extended by (Qian 2008). Qian's definition and fault classification are suitable for systems that are required to produce replicated responses for a given input sequence. Examples of such systems are:

1. TMR (Johnson 1989; von Neumann 1956) systems (see 3.4.1.1 a non-faulty component is required to send its output to three other components or
2. A non-faulty component, which is part of some Byzantine agreement protocol, is required to send its output to other components.

Figure 5.2 shows the response mechanism of a non-faulty component C with multiple r identical outputs, as a result of receiving an input sequence V_i at time T_i.

The resulting outputs are defined as:

1. $V_j = \left\{V_j^1, V_j^2, V_j^3, \ldots, V_j^r\right\}$, where V_j^k, $1 \leq k \leq r$, are r outputs

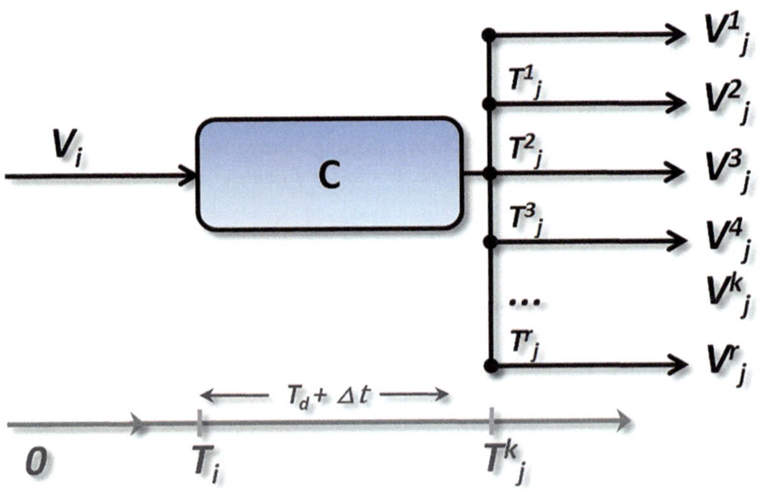

Fig. 5.2 Input-response mechanism of component C with replicated output

2. $T_j = \left\{T_j^1, T_j^2, T_j^3, \ldots, T_j^r\right\}$, where V_j^k, $1 \leq k \leq r$, is produced at time T_j^k.

For a component C, with an input sequence V_i, received at time T_i, its replicated response is correct (correct replicated response) if:

1. $V_j = W_j$, were W_j is the expected vector of replicated outputs; and
2. $T_j^k = T_j = T_i + T_d + \Delta t$ for all k, $1 \leq k \leq r$

Therefore, in a correct replicated response, all individual responses must have the correct values "on time" (according to the specification), but not necessarily "*at the same time*". For instance, responses with same values, which take place at different times, nevertheless within the specified interval, would be part of a correct replicated response.

Such interval is the *skew interval* and it is the period within which all individual responses are produced. It is defined as $\left|T_j^1 - T_j^k\right|$. For any two outputs:

1. $\left|T_j^1 - T_j^k\right| \leq T_{max}$ for all 1, k, $1 \leq 1, k \leq r$.

For replicated-response systems, two types of faults can be considered. A *consistent fault* takes place when individual responses of a component deviate from the specification in an identical manner, whereas *inconsistent faults* are the ones that cause any other breach of the specification.

An incorrect replicated response is a *consistent fault* iff:

1. $V_j^1 = V_j^k$, and $\left|T_j^1 - T_j^k\right| \leq T_{max}$ for all 1, k, $1 \leq 1, k \leq r$.

Note that a consistent fault causes a component to produce identical values (not necessarily correct values) within the skew interval, although "not on time". A few examples of consistent faults are faults with the following properties are:

1. some outputs being on time and the rest are produced early with correct values
2. some outputs being late and having correct values but the rest are correct
3. all outputs having identically incorrect values.

An inconsistent fault is an incorrect replicated response if its individual responses do not satisfy the consistent failure conditions explained above. A *byzantine fault* (the behaviour of a component that gives conflicting values to other components) is a type of inconsistent fault.

5.2.2 *Maintainability: Detectability, Diagnosability and Recoverability*

Fault detection, diagnosis and recovery are required to ensure resilience. Testing and diagnosis may be online, offline or a combination of both (Kaegi-Trachsel et al. n.d.). Online testing and diagnosis are performed concurrently with system operation, whilst offline methods require that the system or subsystem is taken out of service for a specific time.

Often, online testing is used for detection while offline diagnosis locates and identifies the fault(s). As soon as the system/subsystem is repaired, offline testing can be used to verify that the repair was successful before placing it back to normal operation.

Test vectors are used by automatic test pattern generation tools (ATPG) (Agrawal and Chakradhar 1995; Roy et al. 1988a, b) to attempt the detection of all or most modelled fault groups. A *test vector* is a string of n logical values (0,1, or irrelevant X) that are applied to the N corresponding primary inputs (PI) of a circuit, at the same time frame, in order to detect one or more faults (Roy et al. 1988a, b).

The specification of a test vector should have two components: the input to be applied and the expected fault-free output (e.g. $t = I/O = 0010/11$).

A fault will be detected if the output under fault is different than the expected output. If a series of test vectors is applied in a specific order, the term *test sequence* is used, otherwise, it is a *test set* (Roy et al. 1988a, b). The term *test pattern* is often used to refer to any of these three terms.

The testing process involves test pattern *generation* (ATPG), test pattern application in the CUT and output evaluation by an output response analyser (ORA) (Stroud 2002). Figure 5.3 shows the basic testing flow of a circuit, whose output response after processing a test pattern is compared to the expected pattern (fault-free response pattern that a non-faulty circuit would exhibit). The quality of testing will depend on its fault coverage (as defined in previous chapters) and speed.

Following (Abramovici et al. 1994), let x be a random input vector, and $Z(x)$ the function of a circuit under test A with an input x. A fault f would transform A into a new circuit A_f with function $Z_f(x)$. Let be T a test set $T = \{t_1, t_2, t_3 \ldots t_n\}$ formed

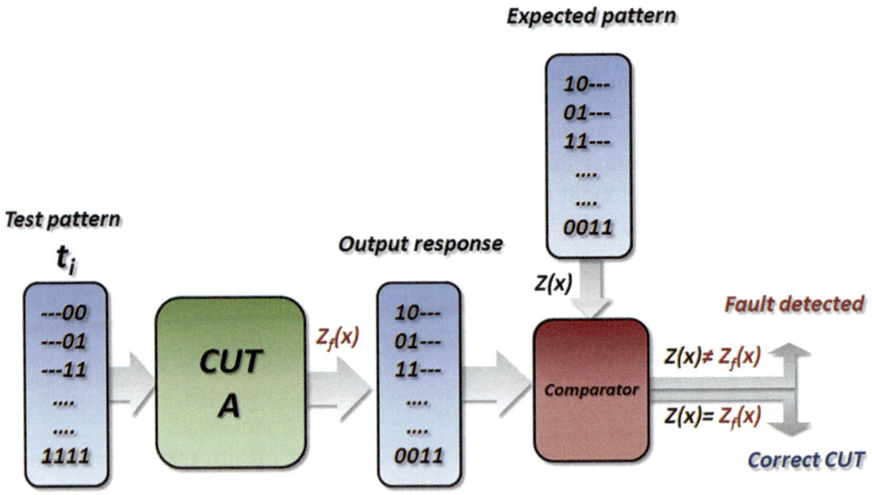

Fig. 5.3 Basic testing flow of circuit under test (CUT)

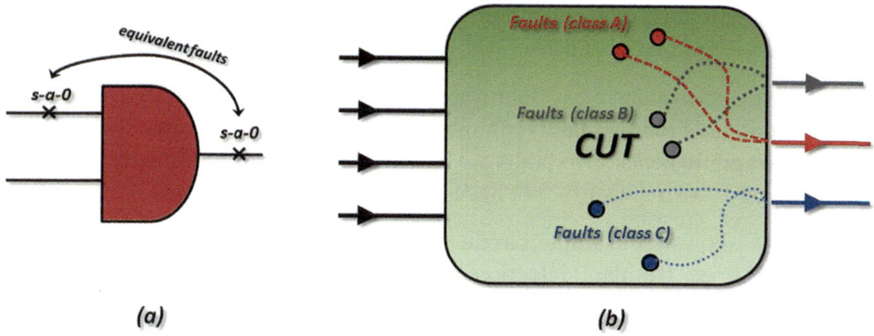

Fig. 5.4 Fault diagnosis and equivalent faults. **a** example of equivalent fault. **b** Fault detection and diagnosis versus vectors

by n t_i test vectors where $t_i \geq 1$. Figure 5.4, the CUT A is tested by applying T and comparing the output response $Z_f(t_i)$ with the expected pattern $Z(t_i)$. A fault is detected if the output response is different than the expected pattern:

$$Z(x) \neq Z_f(x)$$

With regards to fault diagnosis, a test is said to distinguish two faults f_1 and f_2 (*distinguishable/diagnosable faults*) if the output response of the faults are different from each other:

$$Z_{f1}(t) \neq Z_{f2}(t)$$

Conversely, two faults are functionally *equivalent* if all tests that detect f_1 also detect f_2:

$$f_1 \sim f_2 \quad \text{iff} \quad Z_{f1}(t) = Z_{f2}(t) \quad \text{for all } t$$

Functional equivalence can be easily analyzed for logic gates. An example of equivalent faults is shown in Fig. 5.4a. In turn, there is no test that can distinguish between the s-a-0 faults occurring in the input and output of the AND. The same applies to all s-a-1 faults that occur in an OR gate.

Figure 5.4b shows six faults, two of each class A, B and C, in a CUT. Note that detection, diagnosis together with containment and recovery are some of the goals of testing.

For fault detection, at least one vector is needed (fault detection provides only whether the circuit is free of faults or not).

For fault diagnosis, at least one vector that produces different responses for every fault class (fault diagnosis aims to determine time, location and type of the detected fault) is needed.

For combinational circuits with multiple outputs, two types of equivalence can be differentiated (Sandireddy and Agrawal 2005):

1. Diagnostic equivalence: two faults are f_1 and f_2 are diagnostically equivalent if the functions of the two faulty circuits are identical at each output
2. Detection equivalence: two faults are f_1 and f_2 are detection equivalent if all tests that detect f_1 also detect f_2, not necessarily with the same output

Figure 5.5 shows an example circuit with two single *s-a-0* faults in the input *c* and output *Y* lines. Both are detection equivalent faults, but are not diagnostically equivalent.

A fault f_2 dominates f_1 ($f_2 > f_1$) if the test set for f_1 (T_1) is a subset of the test for f_2 (T_2). All tests pattern of f_1 would detect f_2. Therefore, f_1 implies f_2 and including f_1 in the fault list would be sufficient.

If two faults dominate each other, then they are equivalent:

$$f_1 \sim f_2 \quad \text{iff} \quad f_1 > f_2 \text{ and } f_2 > f_1$$

ATPG tools generate test patterns that target possible physical faults according to the fault model (Agrawal and Chakradhar 1995; Roy et al. 1988a, b). An increase in the complexity of circuits involves bigger fault dictionaries and patterns, slowing down the ATPG process.

The implementation of quick detection and diagnostic mechanisms can improve the effectiveness of resilience.

One way of creating compact sets a is fault collapsing, which is the process of reducing the number of faults by using redundancy, equivalence and dominance relationships among faults is called fault collapsing (Abramovici and Breuer 1979). To lessen the burden of test generation, two main types of fault collapsing are used:

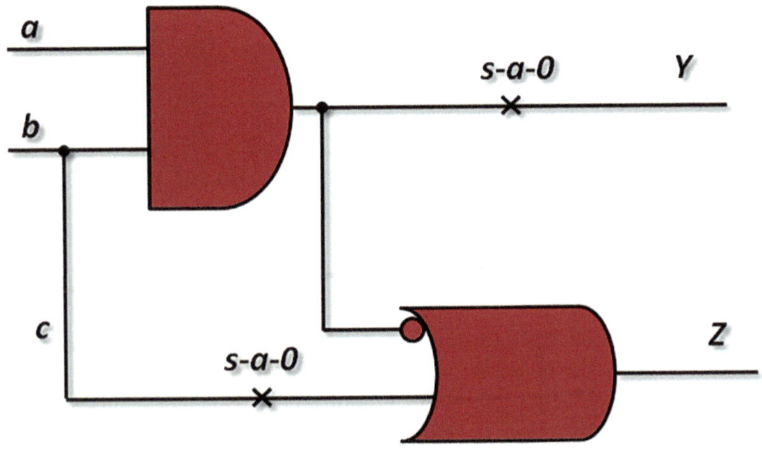

Fig. 5.5 Example of non-diagnostic detection equivalence

1. *Fault equivalence collapsing*: uses the notion of fault equivalence to remove most of the equivalent faults from the pattern. Faults of a logic circuit can be divided into N disjoint equivalence subsets S_i, where all faults within a subset are mutually equivalent. A fault set S_i is *collapsed* if it contains one fault from each equivalence subset
2. *Fault dominance collapsing*: uses the notion of fault dominance to remove dominating faults from the equivalent collapsed faults. If fault f_2 dominates f_1, then f_2 is removed from the fault list

Figure 5.6b shows the dominance and equivalence relationships of a given circuit (Fig. 5.6a).

Both equivalence and dominance relationships are transitive. For instance, Fig. 5.6c—if fault f_2 dominates f_3 and f_3 dominates f_5 then f_2 will dominate f_5. Collapsing algorithms use this transitivity property to reduce the fault patterns (Prasad et al. 2002). If fault detection is the only objective (e.g. fail-safe system that does not require diagnosis), then fault dominance collapsing can be used to further reduce the fault list.

A fault f is *detectable/testable* if there is a test T that is able to test/detect f, otherwise f is an *undetectable/untestable* fault (Agrawal and Chakradhar 1995).

Undetectable faults can be partitioned into two subsets: partially detectable faults and redundant faults. A test set/sequence is said to be *N-detectable* if all faults are detected at least N times with N different test vectors (McCluskey and Tseng 2000). The higher the value of N, the higher is the fault coverage.

A circuit is *redundant* if the function realized by the circuit without fault(s) is the same as the function realized by the circuit with one or more faults (Carter 1979). *Redundant faults* are undetectable faults that do not affect the circuit operation

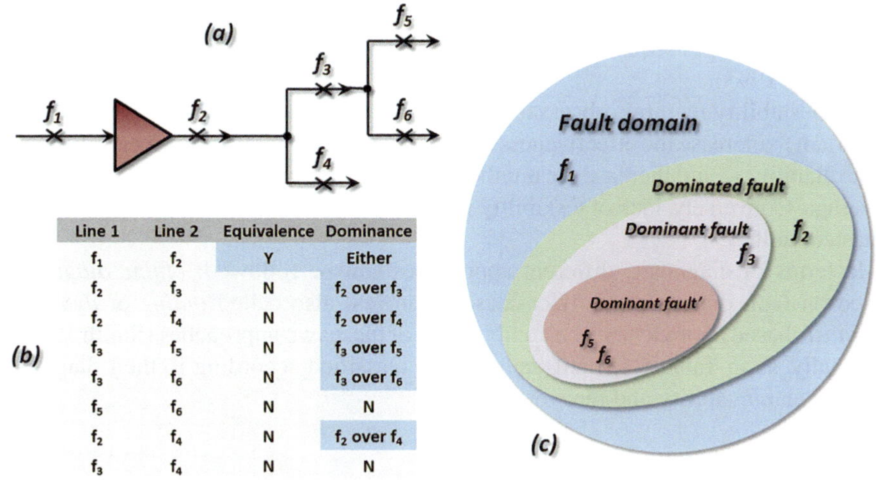

Line 1	Line 2	Equivalence	Dominance
f_1	f_2	Y	Either
f_2	f_3	N	f_2 over f_3
f_2	f_4	N	f_2 over f_4
f_3	f_5	N	f_3 over f_5
f_3	f_6	N	f_3 over f_6
f_5	f_6	N	N
f_2	f_4	N	f_2 over f_4
f_3	f_4	N	N

Fig. 5.6 Example of non-diagnostic detection equivalence

(*operationally redundant*). It can be argued why it is of interest to discover redundant faults if they do not affect circuit logical behaviour.

Discovering and removing redundant faults from the tests is important for the following reasons:

1. In redundant circuits, the presence of undetectable faults can invalidate certain tests, raising problems such as (Friedman 1967):
   a. If f_1 is a detectable fault and f_2 is an undetectable fault, then f can become undetectable in the presence of g. In that case, f_1 is a second-generation redundant fault (Friedman 1967)
   b. If two undetectable single faults g_1 and g_2 are simultaneously present in a system, then they can become detectable
2. Due to time and power consumption constrains, it is not feasible to perform a complete search of all possible faults in any given circuit. There are certain needs for minimizing the tests patterns that detect existing faults.
3. Additional current drains can be induced due to redundant faults such leakage faults (Mao et al. 1990; Xiaoqing et al. 1996) and gate oxide shorts (Hawkins et al. 2003; Segura and Hawkins 2005), which are especially undesirable in low-power devices.
4. Redundant defects may indicate a latent reliability problem.

In the case of combinational circuits, all undetectable faults are redundant faults (Abramovici and Breuer 1979). Testing sequential logic is significantly more difficult that testing combinational logic, whose response is a function of its initial state.

In the case of sequential circuits, it has been shown that certain faults for which a test sequence does not exist, under certain conditions, faulty behaviour may be detected. These are *partially/potentially detectable faults,* faults that affect circuit operation under some states (*partially irredundant*), but are not manifested at the outputs for any input sequence under other states (therefore undetectable) (Pomeranz and Reddy 1993).

The testability features, observability and controllability (defined in Chap. 2) are important to increase the effectiveness of fault detection. Therefore, observability and controllability would increase the number of testable and tested faults and so the fault coverage. Conversely, lack of testability would increase the number of untestable and untested faults.

In terms of diagnosis, different approaches can be followed: *offline diagnosis* based on fault dictionaries, effect-cause diagnosis, also called *online* or *dynamic diagnosis* based approaches, or a combination of these two approaches (Smith 1985).

Finally, as in Table 5.1, faults can also be classified according to their diagnosability, contain–ability and recoverability.

5.3 FT and System Modelling

When a fault appears, extra redundancy is needed to cope with it. Thus, redundancy and ability to use it form a sort of scheme/framework for implementation and use of reconfigurability.

When system and fault modelling are developed together, system behaviour in presence of faults and control process of FT can be considered at earlier stages, taking into account mutual dependencies of solutions at every stage of a process of design and development.

At the same time, redundancy can be used for various purposes and can be involved as essential part of reconfiguration of the connected computers. System reconfiguration purposes are:

- Performance improvement,
- Reliability enforcement,
- Energy-wise use.

The approach of developing reconfigurable computer was named PRE-smart computers (Schagaev 2009; Schagaev et al. 2010). Possible inheritance of properties is shown in Fig. 5.8.

Thus PRE (performance-, reliability- and energy-) wise systems might be designed rigorously, using reconfigurability and recoverability as system features introduced at conceptual level. Success of PRE designs use of this approach for connected computers (networking, clusters multiprocessing) depends on careful trading of the redundancy introduced to achieve the required property.

NB. Why this figure occupies a half of the page? This is because it defines the future of the whole design not only in the ICT.

5.4 Trading P, R, E

Describing a system in terms of *Structure, Information* and *Time* as parameters of redundancy (Schagaev and Buhanova 2001) it is possible to estimate reliability, as well as other properties. Redundancy might be weighted, say, in units or values with or without relation to the steps of any supportive algorithm that applies it to form and control configuration.

While time and information are clear in units—seconds and bits, the structure, especially structural redundancy requires some extra efforts. Note also that time, information and structure are considered as independent variables. Structural redundancy for our purposes might be measured using graph or graph notation:

$$\mathrm{d}S : \langle \mathrm{d}V, \mathrm{d}E \rangle$$

where dS denote introduced structural redundancy, while dV and dE denote extra vertices and edges added in the structure to make step of algorithm implemented.

Then our efforts toward PRE goal can be measured quantitatively as redundancy vector:

$$dR = \langle dT, dS, dI \rangle$$

Time, information and structure as mentioned above are considered as independent variables.

It is clear that reconfigurability of the system can be used for various purposes (Fig. 5.8). To be able to use redundancy and apply reconfigurability we need to consider introduction of supportive tool for reconfigurability—so-called a "syndrome". The syndrome provides a real-time status for every element of the system, it shows a "snapshot-status" of the system in terms of reliability, performance and power awareness.

Let us now analyze how generalized algorithm of fault tolerance might be implemented using mentioned redundancy types and reconfiguration introduced as a system property.

As Figs. 5.7 and 5.8 illustrate, the mentioned mutually contradictive requirements of performance, power consumption and reliability limit design choice in design of embedded systems.

These constrains become even more serious when system is implemented on a chip (SoC). In this case, reliability of elements, performance and power consumption are defined and limited by the same technology.

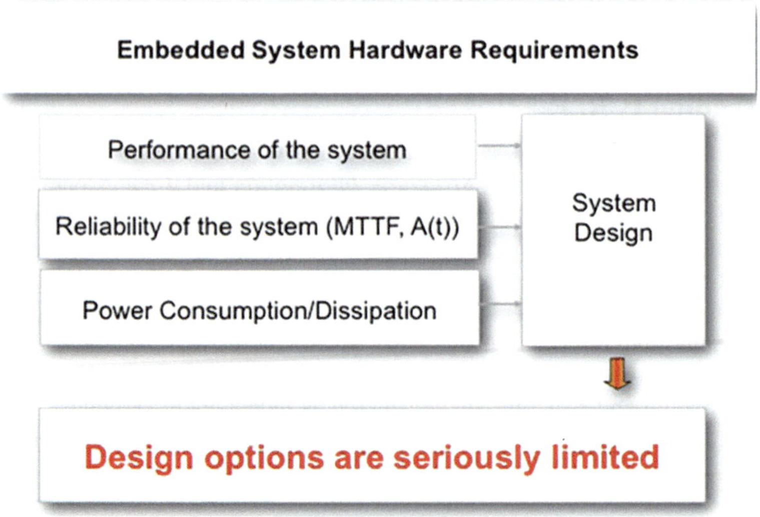

Fig. 5.7 Performance, reliability and power concerns on the design of embedded systems

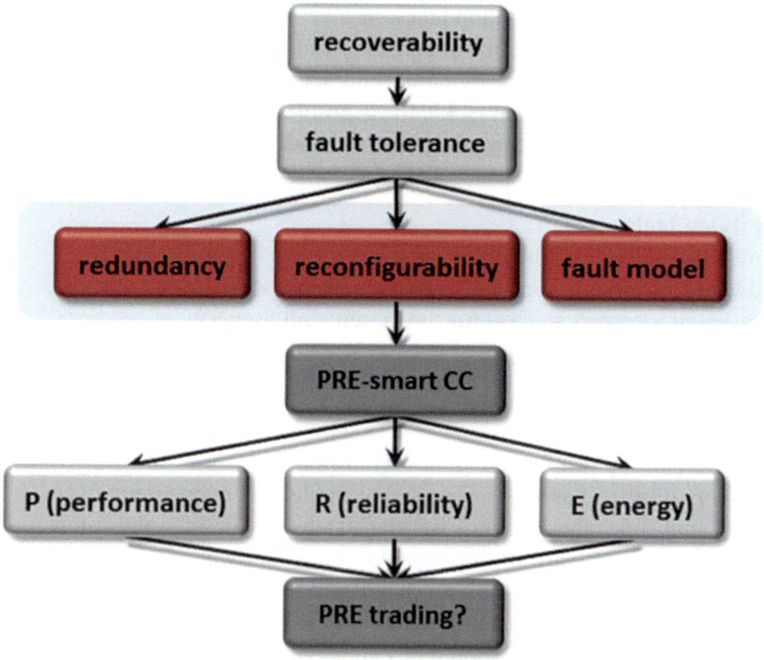

Fig. 5.8 Reconfiguration purposes of a system including fault tolerance

Regarding reliability available redundancy hints a potential solution:

> It is known that maximum reliability of the systems is achieved when all involved parallel elements of the system have equal reliability.

For example, for SoC it technologically means that we need to introduce in the system different redundancy levels for each mentioned group of elements to equalize group reliabilities. This segment of research needs further exploring.

5.5 GAFT: Generalized Algorithm of Fault Tolerance

It is well known that fault tolerance of a computer system can be achieved by introduction of static redundancy in hardware and system software (HW/SSW). It is also well known that using traditional approaches of fault tolerance (Anderson and Lee 1981, p. 81; Avizienis 1971; DeAngelis and Lauro 1976) is expensive in terms of time, information or hardware overheads.

To avoid this, the authors (Schagaev 1986, 1987; Sogomonian and Schagaev 1988) proposed to consider fault tolerance not only as a feature but also as a process that can be implemented algorithmically.

The introduction of static redundancy in HW and SSW might be prohibitively expensive in terms of cost and power consumption. Therefore, a process that implements fault tolerance assuming dynamic interaction of existing redundancy types between elements (Fig. 5.9) can tackle these problems.

Our initial five-step algorithm (Sogomonian and Schagaev 1988) was like this:

- Detecting faults
- Identifying faults
- Identifying faulty components
- Hardware reconfiguration to achieve a repairable state
- Recovery of a correct state(s) for the system and user SW.

Since then, this algorithm has been further developed. The outcome of this work is the Generalized Algorithm of Fault Tolerance (GAFT), a seven-step fault-handling algorithm (see further Fig. 5.9):

The different types of redundancy (information, time and structural either hardware- or software-based) can be used to implement every step of GAFT.

At the same time, there is an essential extension for this algorithm regarding the role of system software in implementation of hardware fault tolerance. Bullet F) of GAFT Fig. 5.11 above has a crucial importance, especially due to latency of hardware faults that is increasing with complexity of computer systems. System software-related part of GAFT should be extended as Fig. 5.10 proposes. For system software, we should find correct state of software first to be able to continue recovery.

Only and when hardware recovery and, if necessary, reconfiguration have been done, the systems software fault-free state should be found and procedure of toleration of fault completed.

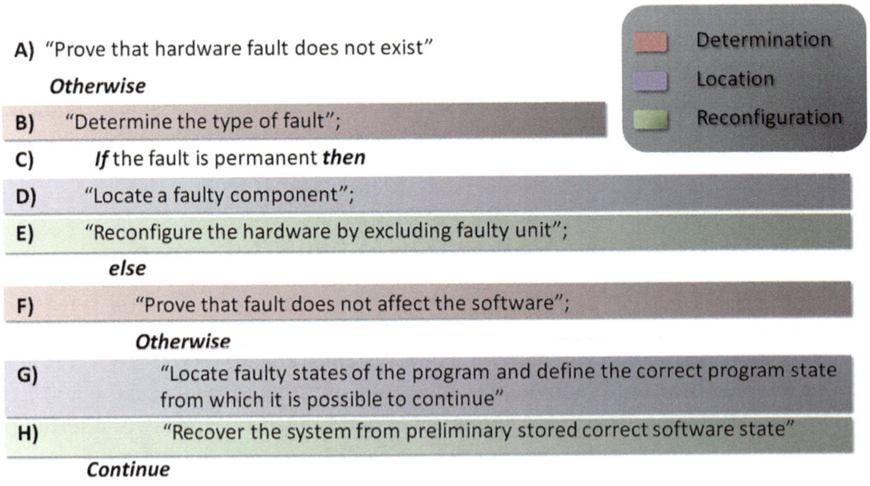

Fig. 5.9 GAFT: generalized algorithm of fault tolerance

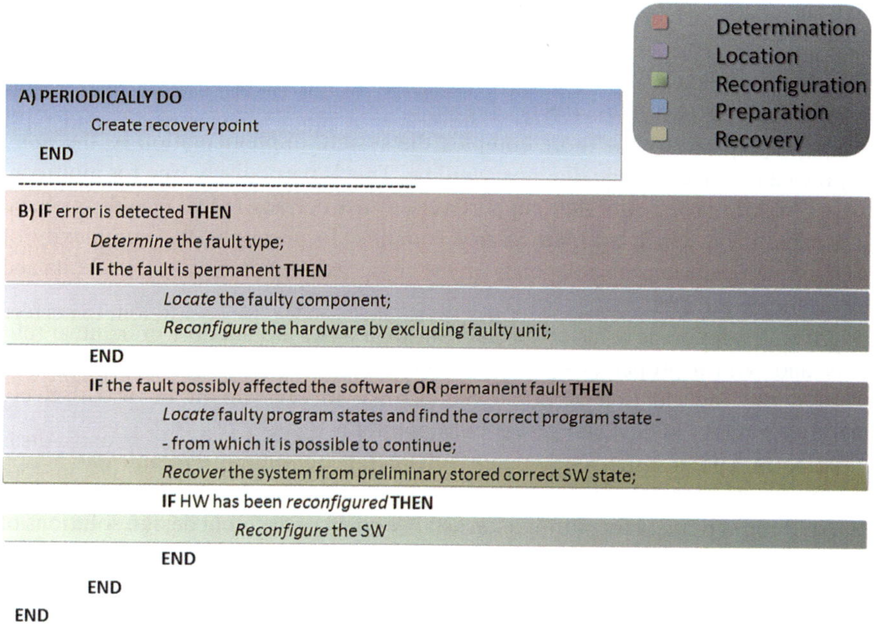

Fig. 5.10 GAFT: generalized algorithm of fault tolerance

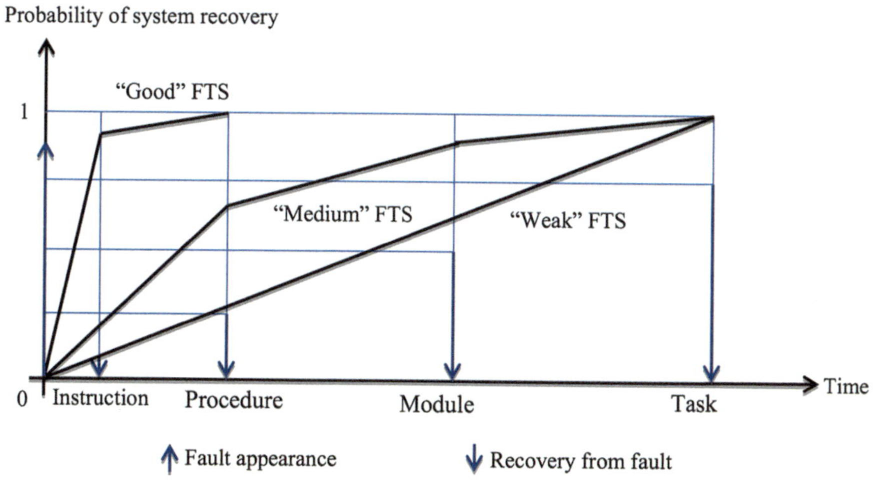

Fig. 5.11 System recovery time according the level of implementation of checking and recovery schemes

When latency is zero, the latest recovery point is sufficient. In the real world we do not know what is size of latency is; therefore, a searching of correct state of software is unavoidable. Fortunately, the problem of finding of software correct state was solved (Schagaev 1986, 1987).

Generally speaking, the more complex the system implementation is, the more complex Fault detection and diagnosis will be. This is particularly true for multicore system (MIMD processors that support vector instructions (SIMD) and pipelined implementations, which are particularly complex. In principle, the complexity of GAFT implementations also depends on the complexity of the system, its faults and fault tolerance models.

Hardware support is a faster mechanism than software support to achieve reliability and with a proper design there should be little performance degradation. However, the introduction of static redundancy in HW might be prohibitively expensive in terms of cost and power consumption.

Therefore, a process that implements fault tolerance assuming dynamic interaction of existing redundancy types between elements can tackle these problems.

GAFT might be used for comparison and overview of different design solutions of FT systems. It also allows controlling fault coverage at every step of system design, providing a tool for the selection of efficient solution and estimation of overheads.

A substantial redundancy can cover multiple steps in GAFT, such as fault detection and fault recovery; take, for example, TMR systems, when a faulty channel output is overruled by two correct outputs.

The implementation of the hardware checking (step A of GAFT) at different levels causes different timing for GAFT completion: microseconds for the instruction level, milliseconds for the procedure level, hundreds of milliseconds for the module level, seconds to tens of seconds at the task level and tens of seconds to minutes or even hours at the system level.

Different implementations would have different properties in terms of timing, fault coverage, types of faults that can be tolerated, power consumption, complexity and cost.

Figure 5.11 illustrates various solutions of fault toleration at the level of instruction, procedure, module and task. All shown systems are fault-tolerant, have different cost, redundancy level involved and coverage of faults tolerated.

It is, therefore, wise to combine different checking and recovery schemes in one system to achieve required specifications, it is wise to combine different checking and recovery schemes in one system.

For example, it might be beneficial to protect the processor and the memory by using hardware schemes at the level of instructions (duplicated processors, triplicate memory) and use higher-level schemes (procedure or module) for the other hardware components due to cost and power constraints.

Processor and memory is used at every instruction execution. The implementation levels are not mutually exclusive; for example, the combination of hardware- and software-based checking can significantly improve fault coverage. In general, the higher the implementation level, the less hardware support is required, but with higher timing and software coding overhead.

A good fault-tolerant system tolerates the vast majority of transient faults within the interval of instruction execution, making them invisible for other instructions (and software).

At the same time, our assumptions about transient "live" fault are arbitrary, thus transient faults with longer time range or permanent faults might be detected and recovered differently, for example, at the procedural or task level of system software.

Taking into account that transient faults occur at an order of magnitude more often than permanent faults, transient fault tolerance must be done extremely effective.

In turn, it is necessary to implement special schemes for HW reconfigurability and recoverability to eliminate the impact of permanent faults on the system.

GAFT completion requires three fundamental processes, called in literature P1, P2 and P3 (Stepanyants and Schagaev 2001).

The error *checking process* P1 is responsible for checking the system state in terms of hardware fault existence/appearance. The second process, the *error recovery process* P2 prepares recovery states when it is scheduled by the system. The third one, the *functional process* P3 is a process of calculations or instruction executions.

When a transient fault is detected, its toleration assumes recovering the information that has been modified and restoring hardware states. The third one, the *functional process* P3 is the process of calculation or instruction execution. Let's assume the primary function of the real-time critical system as "process three" or P3.

Therefore, if the system ensures full functionality and transparent application recovery for the process P3 (from a predefined set of faults in a given time frame) then the system is fault-tolerant.

In other words, *we define a system as fault tolerant if and only if it implements GAFT transparently for applications.*

Or, in a bit more details, as definition:

> If the system ensures full functionality and transparent application recovery for the process P3 (from a predefined set of faults in a given time frame) then the system is fault tolerant.

5.5.1 GAFT: System States and Actions to Implement Fault Tolerance

GAFT above presents an approach of achieving a new and complex property (fault tolerance) considering it as a process with several steps and phases. Further generalization and detailed analysis of system state change are presented below.

At any given time, a single processing element (SPE) system can be in one and one only of five possible states: ideal, faulty, erroneous, degraded and failed. Figure 5.12 shows the five S states, the potential T transitions and the M mechanisms involved in fault tolerance.

Two different areas can be differentiated: the green area at the bottom half represents the conventional environment with no FT capabilities, whereas the red area at the top half represents the possible states and transitions in a dependable environment.

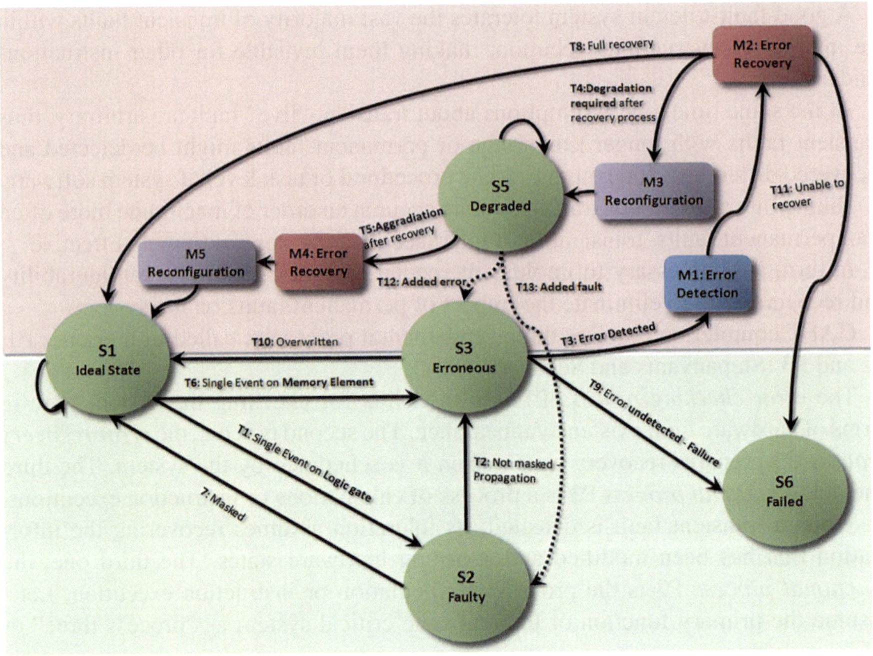

Fig. 5.12 GAFT: generalized algorithm of fault tolerance

Considering S1 as the initial state, a single event can change this ideal state. Note that with the exception of the transition T12, Fig. 5.12 does not contemplate the cases where a single event can occur between transitions.

If the deviation in the form of voltage transient introduced by such event affects combinational logic, the system would turn (Transition 1) into a faulty state S2. The voltage transient may propagate to sequential logic, such as a memory cell or latch, potentially flipping the bit and contaminating the data flowing within the system, and leading the system to an erroneous state S3 (soft error in Fig. 5.12).

However, there are three masking effects that can prevent transition for this particular event from S2 to S3: logical masking, electrical masking, and latch-window masking. Logical masking happens when one of the other inputs of the affected gate is in controlling state so that the output does not vary (e.g., 1 for an AND gate, or 0 for a NAND gate).

Electrical masking happens when the voltage transient impacts successive logic gates and propagates through the logic chain, fading out before reaching the registered output. Latch-window masking happens when the arrival of the pulse is outside the latching window, usually based on the setup time and hold time of the sequential logic.

Nevertheless, if during the ideal state S1, as a result of the single event the voltage transient affects the sequential logic directly, the system state would transit straightforward to an erroneous state S3 (transition 6).

The implementation and the coverage of faults within the system can be measured probabilistically, assuming the existence of undetected faults. We consider that an undetected fault would lead (T9) to failure in the system (S6), unless the error is overwritten [(e.g.: a memory bit that has flipped can potentially be flipped back to the original value by another event before the fault detection mechanisms were activated or before the error leads to a failure) transiting back to an ideal state. The probability of overwritten errors is very small.

The implemented action of fault detection and recovery mechanisms differs in terms of permanent and transient faults. Faults are initially detected by the fault detection mechanisms (M1 in Fig. 5.12).

A detected fault that is not recoverable by the recovery mechanisms (M2) would lead to a failure (T11). In most cases, recovery will be possible in two forms: full recovery (T8) and graceful degradation (T4).

Ideally, a full recovery would turn the system back to the initial state *S1* or, if full recovery is not possible, SSW will make use of the reconfiguration mechanisms (*M3*) to turn into gracefully degraded state S5. In this state, the system can continue operating properly. In some cases, depending on the severity of the failure, the operating quality may decrease. This becomes more obvious if a further fault or error occurs.

Further graceful degradation may be possible depending on the levels of degradation introduced in the implementation. A good example of degradation support for memory can be found in (Bernstein et al. 1993, 1992).

Recovery from a degraded state takes place once the deviation has been corrected. The recovery mechanisms should be able to return the system back to correct state using additional reconfiguration (M5).

Clear, some logic framework, holistic principles to follow might help to design and develop efficient architecture with required properties. Briefly, the next section explains the principles chosen for this work, as it is introduced in (Schagaev et al. 2010), these principles are subject of next chapter.

5.6 Conclusion

1. In this chapter, we have introduced a Classification of faults by manifestation, and generalised algorithm of fault tolerance (GAFT).
2. Regarding fault behaviour in the system, we considered Response-timeliness and Consistency
3. We gave a special attention to maintainability, detectability, diagnosability and recoverability properties of a system.
4. We have introduced the principle of reconfiguration of the system and how to apply it for various design purposes, such as performance, reliability and energy wise-gain, all aiming to improve the efficiency of a system.
5. We have shown that future systems should be able to “trade” properties of performance, reliability and energy efficiency during their life cycle.

6. Additionally, we were able to introduce an element of structural redundancy for the system design. This is, actually, a key to quantify the whole design process with new properties required for the system.
7. Further, we developed our 1988 algorithm of fault tolerance—Generalized Algorithm of Fault Tolerance
8. As further development of generalized algorithm of fault tolerance we have presented coordinated states and actions to achieve fault tolerance of a system, giving a full explanation of system state changes and the actions required.
9. We also classified FT systems as good medium and weak depending on design solutions.

References

Abramovici M, Breuer MA (1979) On redundancy and fault detection in sequential circuits. IEEE Trans Comput 28:864–865

Abramovici M, Breuer MA, Friedman AD (1994) Digital systems testing & testable design, 1st edn. Wiley-IEEE Press

Agrawal VD, Chakradhar ST (1995) Combinational ATPG theorems for identifying untestable faults in sequential circuits. IEEE Trans Comput Aided des Integr Circuits Syst 14:1155–1160

Anderson T, Lee (1981) Fault tolerance: principles and practice. Prentice-Hall

Avizienis A (1971) Faulty-tolerant computing: an overview. Computer 4:5–8

Azadmanesh MH, Kieckhafer RM (2000) Exploiting omissive faults in synchronous approximate agreement. IEEE Trans Comput 49:1031–1042

Bernstein AV, Tomfield YL, Schagaev IV (1992) Storage unit with high reliability characteristics. I. Avtomat. i Telemekh, pp 145–152

Bernstein AV, Tomfield YL, Schagaev IV (1993) RAM of high reliability properties. II. Avtomat. i Telemekh, pp 169–179

Carter WC (1979) Hardware fault tolerance. In: Computing systems reliability. CUP Archive, pp 211–263

DeAngelis D, Lauro JA (1976) Software recovery in the fault-tolerant spaceborne computer. In: Digest sixth international fault-tolerant computing symposium. IEEE Computer Society, Pittsburg, PA, pp 143–148

Friedman AD (1967) Fault detection in redundant circuits. IEEE Trans Electron Comput EC-16:99–100

Hawkins C, Keshavarzi A, Segura J (2003) CMOS IC nanometer technology failure mechanisms. In: Proceedings of the IEEE 2003 custom integrated circuits conference, pp 605–611

Johnson BW (1989) The design and analysis of fault tolerant digital systems. Addison-Wesley

Kadayif I, Sen H, Koyuncu S (2010) Modeling soft errors for data caches and alleviating their effects on data reliability. Microprocess Microsyst 34:200–214

Kaegi-Trachsel T, Schagaev I, Gutknecht J (n.d.) Hardware testing on the level of tasks. In: 30th IFAC workshop on real-time programming and 4th international workshop on real-time software. Presented at the international multiconference on computer science and information technology

Laprie J-C (2008) From dependability to resilience 8, G8–G9

Laprie J-C, Avizienis A, Randell B, Landwehr C (2004) Basic concepts and taxonomy of dependable and secure computing. IEEE Trans Dependable Secure Comput 1:11–33

Mao W, Gulati RK, Goel DK, Ciletti MD (1990) QUIETEST: a quiescent current testing methodology for detecting leakage faults. IEEE international conference on computer-aided design, 1990. ICCAD-90. Digest of technical papers, pp 280–283

McCluskey EJ, Tseng C-W (2000) Stuck-fault tests vs. actual defects. In: International test conference, pp 336–342
Meyer FJ, Pradhan DK (1991) Consensus with dual failure modes. IEEE Transactions on Parallel and Distributed Systems 2, pp 214–222. https://doi.org/10.1109/71.89066
Pomeranz I, Reddy SM (1993) Classification of faults in synchronous sequential circuits. IEEE Trans Comput 42:1066–1077
Prasad AVSS, Agrawal VD, Atre MV (2002) A new algorithm for global fault collapsing into equivalence and dominance sets. In: International test conference. Proceedings, pp 391–397
Qian Y (2008) Information assurance: dependability and security in networked systems. Morgan Kaufmann
Roy RK, Niermann TM, Patel JH, Abraham JA, Saleh RA (1988) Compaction of ATPG-generated test sequences for sequential circuits. In: IEEE international conference on computer-aided design, 1988. ICCAD-88. Digest of technical papers, pp 382 –385
Roy RK, Niermann TM, Patel JH, Abraham JA, Saleh RA (1988) Compaction of ATPG-generated test sequences for sequential circuits. In: IEEE international conference on computer-aided Design, 1988. ICCAD-88. Digest of technical papers, pp 382–385
Sandireddy RKKR, Agrawal VD (2005) Diagnostic and detection fault collapsing for multiple output circuits. In: Design, automation and test in Europe, 2005. Proceedings, vol 2, pp 1014–1019
Schagaev I (1987) Algorithms to restoring a computing process. IBID
Schagaev I (1989) Yet another approach to classification of redundancy. In: Proceedings of FTSD, Prague, Czeschoslovakia, pp 485–490
Schagaev I (2009) ERA: embedded reconfigurable architecture—past present and future
Schagaev I, Buhanova G (2001) Comparative study of fault tolerant RAM structures. In: IEEE DSN01, Goteborg
Schagaev I, Kaegi T, Gutknetch J (2010) ERA: evolving reconfigurable architecture. In: Proceedings of 11th ACIS. Presented at the international conference on software engineering artificial intelligence, networking and parallel/distributed computing, London
Schagaev I (1986) Algorithms of computation recovery. IBID, p 7
Segura J, Hawkins CF (2005) Bridging defects. In: CMOS electronics. Wiley, pp 199–222
Smith GL (1985) Models for delay faults based on paths. In: Proceedings of the international test conference, pp 342–349
Sogomonian E, Schagaev I (1988) Hardware and software fault tolerance of computer systems. Avtomatika i Telemekhanika, pp 3–39.
Stepanyants A, Schagaev I (2001) Fault tolerant processor and its reliability analysis. In: IEEE DSN01, Goteborg
Stroud CE (2002) A designer's guide to built-in self-test. Springer
Thambidurai P, Park Y (1988) Interactive consistency with multiple failure modes. In: Proceedings of the 7th symposium on reliable distributed systems, pp 93–100
Von Neumann J (1956) Probabilistic logics and the synthesis of reliable organisms from unreliable components. In: Automata studies. Princeton University Press, pp 43–98
Weaver C, Emer J, Mukherjee SS, Reinhardt SK (2004) Techniques to reduce the soft error rate of a high-performance microprocessor. In: Proceedings of the 31st annual international symposium on computer architecture, pp 264–275
Xiaoqing W, Saluja KK, Kinoshita K, Tamamoto H (1996) Equivalence fault collapsing for transistor leakage faults. In: IEEE international workshop on IDDQ testing, pp 79–83

Chapter 6
Hardware Support of Resilience

Abstract The chapter presents structural design of resilient architecture, describing all three areas of data handling: active zone where data processed, interfacing zone where data transfer between architecture elements, and passive—where data stored. Processor as it should be for resilient architectures is presented in full details—hardware schemes, instruction set, comparison with other processors used in the same area. Interfacing zone described with new communicative element so-called T-logic that "glue system together". Passive zone, memory that is designed with extreme flexibility to fit all demands: reliability of a system, performance and energy-saving. New system hardware element to manage resilience of functioning so called a *Syndrome* is introduced and explained. Its functioning within the system in concert with a new processor that best fit resilience of a system explained. Drawbacks of existing processors that market tries to procrastinate into the areas where safety criticality of functioning are discussed. Complete reconfiguration scenario for a proposed computer system is considered, explained and tested. Prototype system with required features of resilience is presented and explained.

Keywords Resilience · Reconfiguration · Reliability · Reconfiguration · Processor · Memory · Interface logic · Resilient architecture

6.1 Introduction

So far, in previous chapters, physics of damages, models of reliability of element and systems were discussed, phenomena of damages causing a failure of the system were also covered.

Element level of a system was be taken into account. We were extracting, where possible, features of hardware elements that suit good for implementation of resilience.

While memory damages causing data errors were discussed in terms of error detection and error correction codes, the system level of hardware and internal structural support of resilience and fault tolerance was not discussed so far.

© The Author(s), under exclusive license to Springer Nature Switzerland AG 2026

I. Schagaev, *Resilience of Computer Systems*,
https://doi.org/10.1007/978-3-032-00938-8_6

At the same time, thinking about resilience, fault tolerance, one has to consider also an architectural, structural solutions for our systems to match non functional requirements of the system.

Thus, this chapter is an attempt to address how to close the gap in terms of system resilience at system and element levels.

Any new complex property, such as resilience should be considered as a process. In this chapter, we read reconfiguration properties of a computer system, considering it also as a process. This process should enable system "fine tuning" for various applications when reliability-, performance- and power-wise functioning is required—all match requirements of resilience.

We present the necessary and detailed argumentation of the system configuration, also covering our processor, deliberately designed to work within resilient systems. Detailed analysis of processor, including detailed comparison with known architectures presents pros and cons of special processor approach.

We also introduce the concept of system syndrome. Its structural design, its structure and its operations are described using a prototype of evolving reconfigurable architecture as an example.

Detailed explanations of use and implementation of the syndrome is given for generalized algorithm of fault tolerance.

We present a hybrid HW-SSW co-design approach and architecture with the ability to reconfigure, satisfying various changeable requirements in performance, reliability and power consumption, required in different environments.

In today's multi-business-use scenario of ICT, the introduced architecture has advantages to the known architectures and has a good compromise between area, performance, power, cost and efficiency of the system implementation.

Some kind of structure of this chapter is presented in Fig. 6.1.

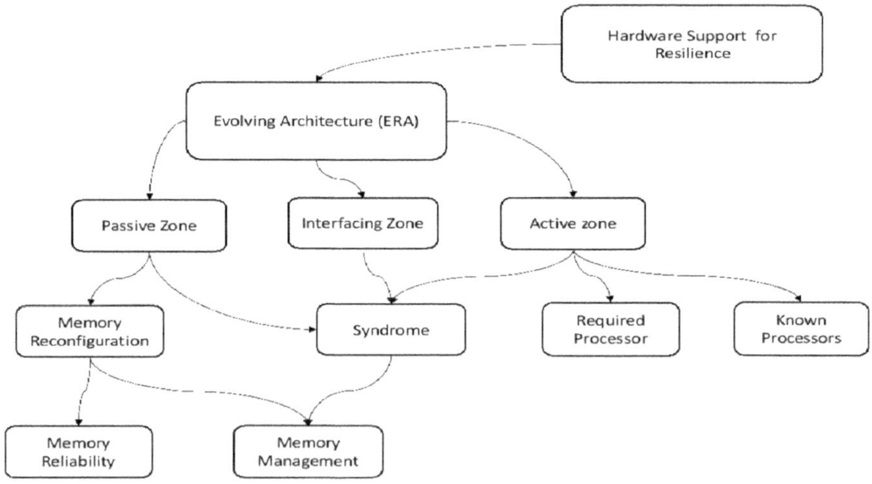

Fig. 6.1 Structure of chapter

6.2 ERA Concept, System Design and Hardware Elements

To develop a new architecture that is almost free from previous drawbacks we proposed for ourselves several "holistic" principles. "We" here mean Niklaus Wirth, Juerg Gutknecht and Igor Schagaev—our discussions since 1989 were actually about these principles as well. We were using these principles for primarily for the ERA (Evolving, Reconfigurable Architecture (Schagaev et al. 2010).

This principle means that, considering through the whole process of design and development the following principles should be pursued:

simplicity, redundancy, reconfigurability, scalability, reliability and ***resource awareness.***

Following these principles, a new hardware architecture (HW) and system software (SSW) have been developed. Brief introduction of these principles follows:

Simplicity: Complexity is difficult to implement and handle efficiently. In addition, big complex systems are more prone to faults, thus lowering reliability.

Reliability: The highest reliability of individual components is preferable but always keeping in mind the cost-efficiency of its implementation.

Redundancy: Deliberate introduction of hardware and software redundancy provides the required level of reconfigurability to reach performance and reliability goals.

Reconfigurability: Apart from the simplicity, reliability and deliberate introduction of redundancy, it is essential to achieve balance between performance, reliability and power. Reconfigurability serves three main purposes: performance, reliability and power awareness. It allows the system to adapt twofold: 1st to recover from a permanent fault and 2nd adjusting the requirements of the running application.

Scalability: Scalability should be kept in mind when designing a system so that it can be extended if the requirements change.

Power-awareness: Mission critical systems have significant limitations of hardware resources and power consumptions constraints (e.g. battery life). Thus, for wise resource use, reconfigurability must be introduced.

Following these principles, the processes of design and development of new architecture can be defined better. It did work for our architectures and systems, thus "field tests" were successful. Let us see how they are applied for new architecture we propose.

Figure 6.2 shows a computer system as three semantically different (from point of view of information processing/transformation) elements. Above-mentioned principles might be applied at the level of each element enabling design by far more efficient computer system.

In terms of information processing point of view the hardware is based on a Single Processing Element (*SPE*) that is divided into three areas (Fig. 6.2):

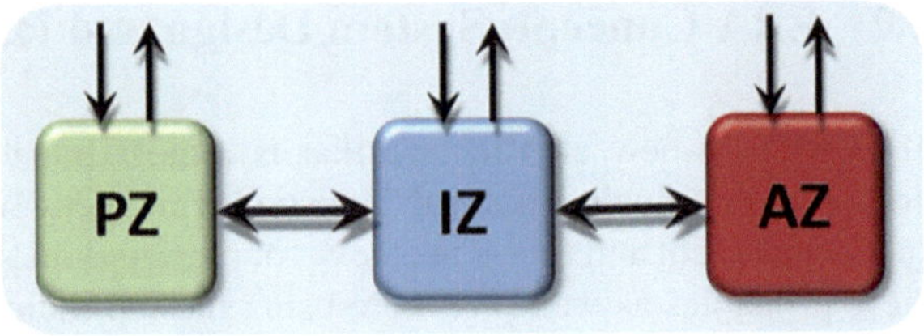

Fig. 6.2 System zones from a information processing point of view

1. At first, the information transformation area—further called active zone (*AZ*) and
2. At second the information storage area—called passive zone (*PZ*)
3. The interconnection of these zones is the interfacing zone (*IZ*).

Regarding information processing all three zones do have different functions, and, therefore, properties and need, therefore, different redundancy mechanisms to tolerate faults and make system reconfigurability possible and efficient.

In brief, some elaboration of each zone:

Active Zone: The active zone consists of the arithmetic unit and logic unit, both separated for better fault isolation and easier implementation of hardware tests.

Interfacing Zone: This includes all communication components such as the processor, the memory buses and the reconfiguration logic. A configurable bus allows the reconfiguration of the hardware to exclude failed hardware components and go into a degraded state, or replace the failed component with a working one.

Passive Zone: This includes basic storage systems, such as memory, that do not act by themselves but are handled by controllers or devices.

The first proposed structure with reconfiguration to deal with each zone is shown in Fig. 6.3.

Minimum deliberate redundancy has been introduced in the form of buffer, register files, replicated memory modules, majority schemes (in terms of HW) and interfacing logic.

System software extra elements required to support fault tolerance are: checkpoint monitor, recovery point monitor, process synchronization and reconfiguration monitors. They named monitors to express their uninterruptable mode of operation. Full details of new system software for resilient, fault-tolerant computer systems have been described by us several times over 20 years, the newest and most comprehensive is given in (Schagaev and Gutknecht 2004).

As mentioned, hardware can be considered as three zones (Figs. 6.2 and 6.3). All elements in these zones have to be reconfigurable for their own purposes as well as other zones requests.

All zones might have different reconfiguration properties. Reconfiguration might have internal and external reasons. For example, when the system forms a configuration for a task execution, it might deliberately and *externally* exclude some hardware

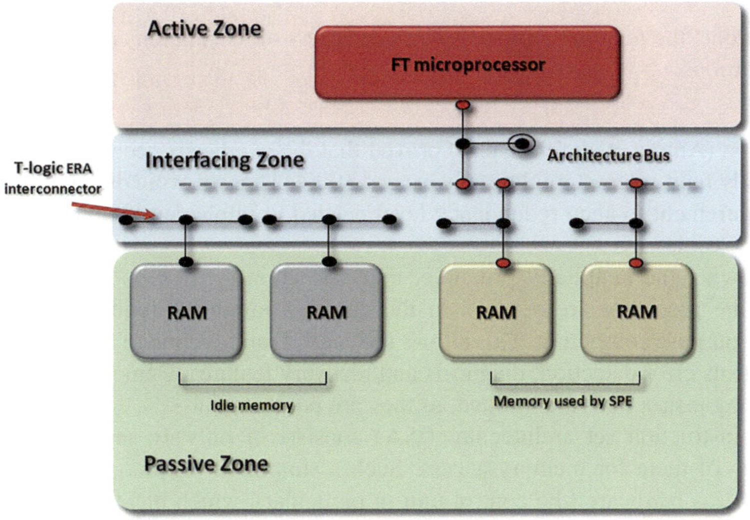

Fig. 6.3 System zones from a information processing in ERA

elements from configuration due to task requirements or transient/permanent fault. In turn, checking schemes activates *internal* reasons for reconfiguration.

Interactions between zones define the level of reconfigurability and flexibility of the architecture. These new hardware reconfiguration abilities must be reflected and supported as new features of the architecture.

6.3 ERA Hardware Configuration: ERA

6.3.1 Active Zone

The main principle used in the design of the processor is simplicity. The cost of P1 and P2 implementation of GAFT depends on the structure of the processor and might become prohibitively high. For readers of this chapter only, we present a cut from our previous works explaining used here abbreviations:

GAFT (generalised algorithm of fault tolerance)—completion requires three fundamental processes, called in our previous books as P1, P2 and P3. The error *checking process* P1 is responsible for checking the system state in terms of hardware fault existence/appearance. The second process, the *error recovery process* P2 prepares recovery states when it is scheduled by the system. The third one, the *functional process* P3 is a process of calculations or instruction executions.

Antola (Antola et al. 1986, p. 1) proved that the overheads necessary to make a CPU fully fault-tolerant might easily exceed 100%. Thus, to avoid duplication, there is a requirement to keep redundancy level needed to implement fault tolerance, as low as possible.

Following the simplicity principle, the instruction set and its implementation within the processor are reduced to the absolute minimum required to support general-purpose computing. This allows the careful introduction of redundancy that implements error detection, diagnosis and recovery features. Complicated memory addressing instructions are omitted, as they are not essential.

The instruction set architecture (ISA) consists of only 16 instructions, with only two of them for memory access. Such a simplified instruction set generally requires less hardware (the control unit in particular), which increases, by design, the reliability of a single processor.

Performance might also be increased as operating clock frequencies can be made higher. Extra details on the instruction architecture are explained further in special chapter.

The proposed microprocessor design offers clear advantages in comparison to CISC architectures: fewer and simple addressing modes, hardwired design (no microcode), fixed and simple instruction formats. Note that simple instruction format allows fetching of two 16-bit instructions per machine cycle. The execution steps are similar to other RISC processors.

There is no pipelining and all steps of one single instruction are executed within one memory cycle. Pipelines are one of the most vulnerable elements of modern microprocessors. The relative amount of area of the chip dedicated to pipelines is increasing with scaling design complexities. For instance, instructions can stall in the instruction queue and the longer they reside there, the higher the chances of getting struck by an energetic particle.

A transient fault in a latch or a memory cell within the pipeline (e.g. SET or SEU) can propagate and become an error at the micro-architectural level (e.g. Register file or Instruction Register). Consequently, the effects of ionizing radiation in this area can lead to SEFI's, severely decreasing the overall reliability.

The absence of pipelining or caches greatly simplifies the processor design, which, in turn, simplifies the implementation of fault tolerance. A careful introduction of redundancy for checking and recovery allows the processor:

1. To detect transient faults during the execution of the instruction
2. To abort the current instruction and
3. To re-execute it, all transparently to software.

There are known arguments that simple ISA's do not have enough instructions to perform most of the application operations. These argue that a small ISA would result in higher compilation effort/time, and that the resultant programs would be bigger, which would increase the amount of memory needed.

However, complex functions are well handled by the compiler instead of having specialized instructions within the ISA dedicated to very particular tasks. Regarding the size of the programs, it is worth mentioning that:

1. A bigger ISA, such as CISC involves longer operation codes which, in turn, increase the size of programs
2. In an architecture based on a smaller ISA, register references require fewer bits
3. It is usually claimed that smaller ISA require more memory. ERA architecture has unique design property: size of all instructions is 16bit while size of a word is 32. Thus, the instruction density per word is 2 and the previous arguments above about bigger code size for RISC do not stand to ERA
4. Since the price of the memory is very low and keeps decreasing this argument also becomes less important.

The processor has a large register file with 32 general purpose registers with a width of 32 bits and no restrictions on their use, which simplifies software development (see Fig. 6.4).

All standard instructions expect exactly two arbitrary registers as input, and save the result of the operation in one of these two registers, thus overwriting one of the input values.

The impact of the size of the register file on overall performance of processor is a question of further research.

Memory access is only possible for 32-bit words at a time, and it has to be aligned. The main structure of the processor architecture is illustrated in Fig. 6.4.

The instructions are fetched from memory into the instruction register and decoded by the Control Unit, which also manages the execution of each instruction. Operands for each instruction being executed are fetched from either the register file or memory multiplexed into the Arithmetic or Logic Units, AU and LU, respectively.

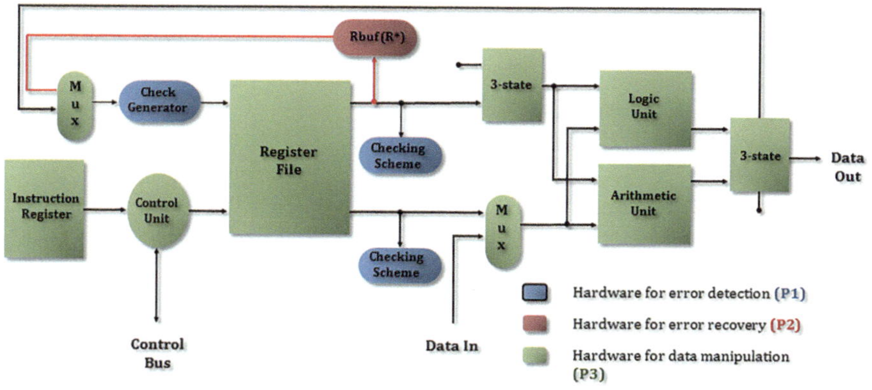

Fig. 6.4 Architecture of active zone

The output data from AU or LU goes either to memory through the data bus or is written back to the Register File. The current value and type of the data might also indicate an address for branch instructions.

The hardware for fault detection and error recovery process P1 and P2 are marked in blue and red, respectively. The hardware for the data manipulating process P3 is marked green.

The simplified architecture Fig. 6.4 presents allows implementation of GAFT at instruction level with reasonably small reliability (13%) (Schagaev 2008).

As before, the three processes are essential for the guaranteed and successful execution of each instruction. Two processes P1 and P2 cope with fault checking and recovering of transient errors, respectively.

P1 is initialized at the start of the instruction execution; P2 is required and initiated when fault has been detected, but it is essential that the pre-modified state is stored at the start of execution of every instruction.

P1 and P3 can operate concurrently. P1 and P2 have an influence on each other: the higher the fault detection coverage achieved by P1, the more successful recovery should be (Stepanyants and Schagaev 2001).

When data is written into the Register File, the Check Generator (marked in blue in Figs. 6.4 and 6.5) generates the checking information for the 32-bits data storing into the register file; this information allows the stored data to be verified later on.

The Checking schemes (marked in blue in Figs. 6.4 and 6.5) check the data integrity when data are read out from the Register File and, when it is possible; correct the data before the ALU operation takes place.

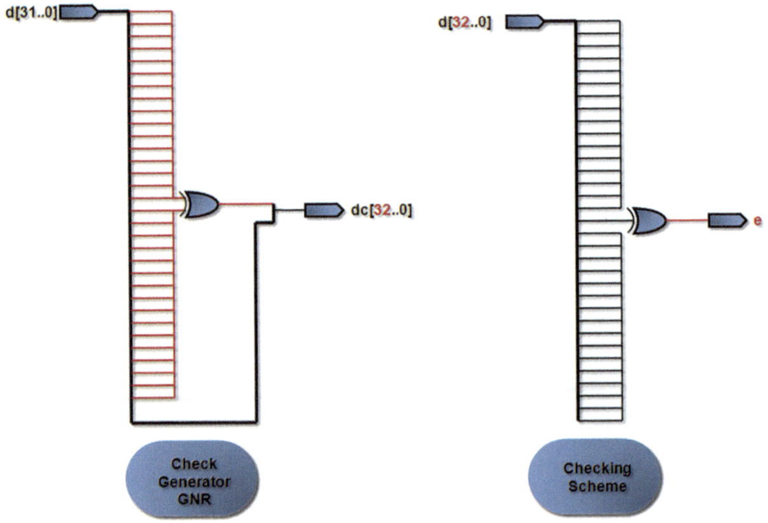

Fig. 6.5 Check generator and checking scheme

In order to implement the fault recovery process P2 an extra Register Buffer Rbuf(R*) is introduced (marked in red in Fig. 6.4). The register buffer is allocated to keep pre-modified state of operand for the currently executed instruction.

6.3.2 ERA Processor Architecture

Here we discuss essential features and solutions made in the ERA design. We start with the needs of detection and recovery going further to Instruction format and designs required.

When a fault is detected during instruction execution, it allows the processor to restore to the initial state before the execution of the instruction enabling the instruction to be repeated. This enables us to tolerate faults within instruction execution.

Here we introduce the extra Register Rbuf(R*), the checking schemes and the reverse instruction sequencer combine to make the implementation of P1 and P2 possible without any perceptible time overheads.

Having this kind of overheads in mind we strive to keep the redundancy level as low as possible and still needed to achieve fault tolerance. In comparison to other architectures such as complex instruction set computer (CISC), ERA differs in the following: one addressing mode, hardwired design (no microcode), regular instruction set, few and simple instructions. The difference of ERA to RISC processors is not as big as to CISC processors, but the number of instructions is still seriously smaller and the instruction complexity is reduced.

Figure 6.6 illustrates a simplified version of the instruction execution of the ERA processor. The execution steps correspond to the ones in other RISC processors, with the exception that no pipelining is used.

Therefore, all steps of one instruction are executed within one processor cycle and the special used memory controller is used only when special instruction stores result in memory.

We briefly discuss the steps involved in the execution of one instruction:

1. Fetch Instruction: This step loads an instruction from main memory and stores it in the processor internal instruction register (IR). As every instruction is only 16-bit wide (see further sections), this step is needed every second instruction if the instruction is not a jump instruction.
2. Decode Instruction Decodes the instruction.
3. Execute Instruction: Executes the instruction, i.e. calculates the result of the instruction.
4. Store Result: The instruction execution might have affected register content, processor flags and other processor state, which is updated and written back in this last step.

At first, it could be argued that the proposed instruction set architecture—further ISA—does not have enough instructions to perform most of the operations, that such

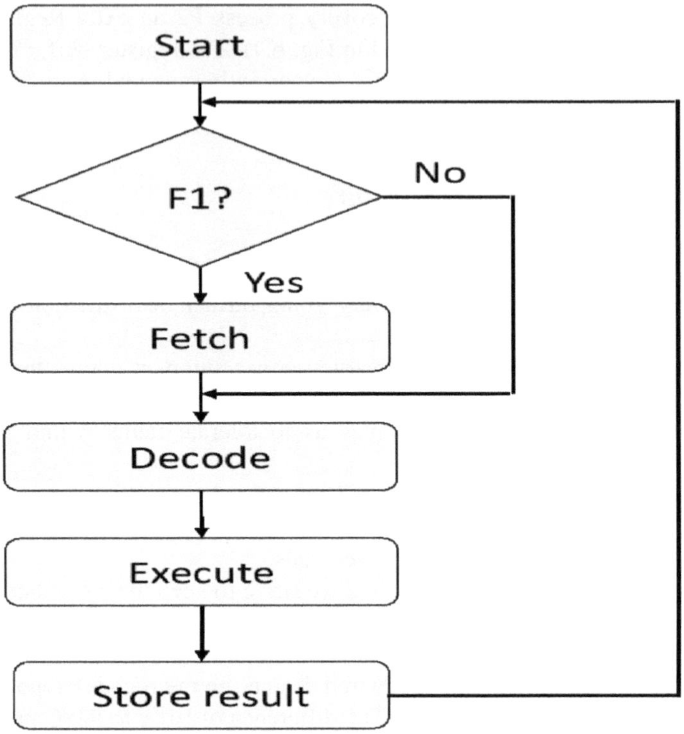

Fig. 6.6 ERA instruction execution—simple version

a small ISA would result in higher compilation effort/time and that the resulting programs would need more space.

However, complicated functions are best handled by the compiler instead of having specialized instructions within the ISA dedicated to very particular tasks. The program size of applications compiled for ERA are, indeed, often larger than the same application compiled for other architectures.

The 16-bit instructions help to keep code size small, but the lack of direct immediate support in instructions increases code size significantly. The lack of some often-used instructions, such as multiplications, also increases code size, as the corresponding library calls need additional setup code. However, we consider this as acceptable to keep the processor architecture simple and be able to guarantee recoverability of every instruction execution.

The processor has a large register file with 32-bit wide general-purpose registers. All normal instructions expect exactly two input registers and save the result of the operation in the second register, therefore always overwriting the second operand content.

Memory access is only possible for 32 bits at a time and has to be aligned. The ERA has internal states (F1, F2), which are required by the special memory controller. The main philosophy of the ERA architecture is to keep it simple; this principle procrastinates also to the memory controller.

The processor and memory controller are designed not to stall the processor due to pending memory operation. This is achieved by interleaving instruction fetch and memory operations. All ERA instructions are 16-bit wide; an instruction fetch operation therefore always loads two instructions from memory and stores them in a processor internal instruction register (IR).

In sequential instruction execution, the compiler can schedule memory instructions in every second instruction slot where no instruction must be loaded by the processor. This concept is illustrated in Fig. 6.7, an extended version of Fig. 6.6 that includes memory operations.

In state F1, the processor fetches an instruction from memory, and the memory unit is therefore busy in this cycle and cannot execute a memory operation at the same time. If the processor is in state F2 (not in state F1), the processor can execute a memory instruction. The compiler is responsible to schedule the instructions in proper order.

If the executed instruction is a branch, the processor switches automatically to the F1 state, as the memory controller can only load 32-bit aligned addresses. Jump

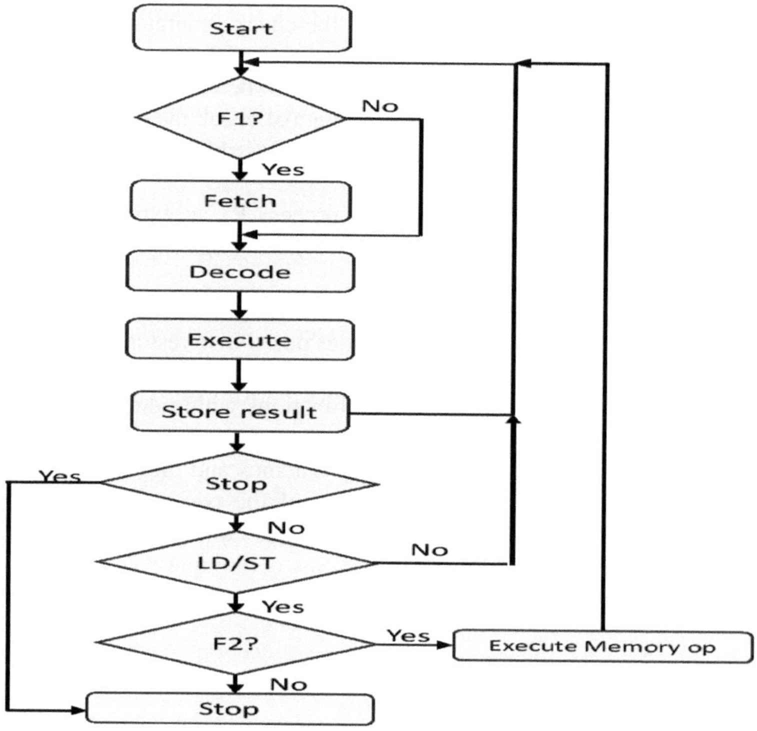

Fig. 6.7 ERA instruction execution—extended version

destination locations must therefore also be 32-bit aligned. All these requirements force the compiler to fill gaps with NOP operations.

The main structure of the processor architecture is shown in Fig. 6.4. The instructions are fetched from memory into the instruction register and decoded by the Control Unit, which also manages the execution of each instruction.

The operands for each executed instruction are fetched from either the register file or memory, multiplexed into the arithmetic and logic units (AL and AU) for operation. The output data from both goes either to memory through the data bus or is written back to the register file. The current value and type of the data might also indicate an address for branch instructions.

The simplified architecture presented in Fig. 6.4 allows implementation of GAFT at the instruction level. As before, the three processes are essential for the guaranteed and successful execution of each instruction.

Two processes P1 and P2 (blue and red), deal with fault checking and recovering of malfunctions, respectively.

P1 is initialized at the start of the instruction execution; P2 is required and initiated when a fault has been detected, but it is essential that the pre-modified state is stored at the start of execution of every instruction. P1 and P3 (actual processing) can operate concurrently. P1 and P2 have an influence on each other: the higher the fault detection coverage by P1, the more successful recovery gain is possible.

When data is written into the register file, the check generator (marked in blue in Fig. 6.4) generates the checking information for the 32-bits data stored into the register file; this information allows the stored data to be verified later.

The checking schemes (marked in blue in Fig. 6.4 check the data integrity when data is read out from the register file and when it is possible, correct the data before the AU or LU operation takes place.

In order to implement the fault recovery process P2 an extra Register Buffer Rbuf(R*) is introduced (marked in red in Fig. 6.4). The register buffer keeps record of the modifying state of the processor.

When a fault is detected during instruction execution and before fault propagates through the architecture further it becomes possible to restore the processor to the initial state. This scheme of instruction execution is enabling the instruction to be repeated and therefore to tolerate the malfunctions within the instruction level, invisible for system and application software.

The extra register Rbuf(R*), the checking schemes and the reverse instruction sequencer combined allow the implementation of the two processes P1 and P2 without any perceptible time overheads. Note that this current version covers single event upset (SEU) only and must be modified in case of multiple events' upset (MEU).

6.3.3 Instruction Set

All the requirements to the processor, which are described in (reference) can only be achieved by using an ultra-reduced instruction set, consisting of 16 instructions where only two of them are memory instructions (load/store).

This allows to save the state of the processor before each instruction execution, and, if a fault is detected, to cancel the instruction and re-execute it on the same processor.

Furthermore, such an instruction set generally requires less hardware, which increases the reliability of a single processor.

The processor has 32 general-purpose registers. They have a width of 32 bits and are equal, i.e. all registers can be used for every instruction. The instructions are designed as two register instructions, where the second register is used to store the result.

The compiler designer must keep in mind that the content of the second register is overwritten. Memory access is 32-bit only and has to be aligned to 32-bit.

Every instruction is encoded into 16 bits (see Fig. 6.8). The two highest bits (bits 14, 15) indicate the format of the operation, which can be 8, 16 or 32 bits.

Not all instructions support a format other than 32 bits, but if they do, they only use the specified number of bits in their operand registers and ignore the others. Bits 10–13 contain the instruction opcode.

Bits 5–9 identify the first operand (register) and bits 0–4 the second operand.

The first and second operands can be any of the 32 general-purpose registers. Constants cannot be directly encoded in the instructions.

Opcode 0 is a special case and is used for four different instructions.

Depending on the format, it can indicate a STOP, a NOP, a debug TRACE or RETI, which is used to exit an interrupt handler.

Another special case is the instruction LDA, which loads a constant to the specified register. The constant is loaded from the next aligned 32 bits after the active program counter.

The program counter then skips the constant and continues the program execution directly after the constant.

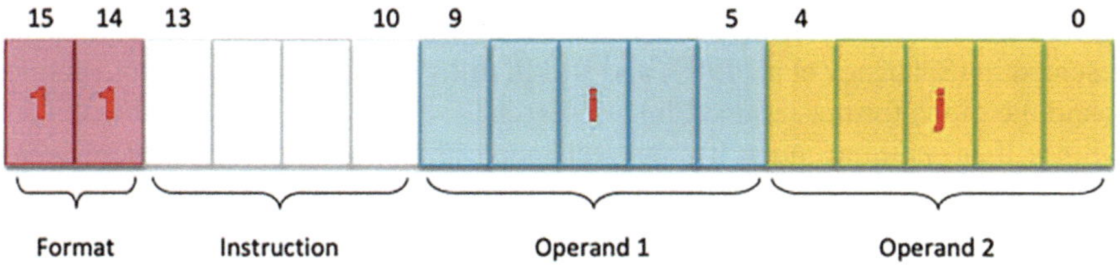

Fig. 6.8 Instruction set format

In general, placing two instructions into one 32-bit word increases code density and performance. The processor executes the left instruction first and can then, without accessing the memory, directly execute the second one.

Note that the program counter has the same value for both instructions. This can be a problem because it is obviously not possible to jump to an instruction on the right side, as it has no unique address and is not 32-bit aligned.

To avoid this, a compiler has to make sure that NOPs are inserted at the correct places to prevent these cases if instruction reordering fails to fill the gap.

All arithmetic operations treat the value in the operand as a signed values. There is no method execute unsigned operations.

Except for the conditional jump instruction every operation accepts the same register for both arguments.

Table 6.1 lists all available instructions together with a short description of their effect.

While most entries are self-explanatory, the result of the compare instruction CND requires some elaboration. Like other platforms, it checks for three conditions, *greater, smaller* and *equal* and saves them as flags in the first three bits of the result register, Table 6.2.

Using an appropriate constant mask and the bit instructions, the result for every possible comparison operation can be extracted from the flags and saved as a Boolean value or used as an argument in a conditional jump, Table 6.3.

6.4 Architecture Comparison and Evaluation

Some notes for architecture comparison. Currently, there are several processor architectures predominates in the world; starting still from × 86 architecture on the desktop and the ARM in embedded devices.

Sure, and therefore they and others should be included in an architectural comparison. Even a brief analysis of these architectures, based on Intel (2009, 2004), ARM (2005), Hennesy and Petterson (2002), Verlag (2009), SPARC (1992), Schagaev et al. (1989), LEON3 (2009) should be enough to justify our own efforts to develop own ERA that fit resilient and reliability way better.

The core of analysis of instruction set and impact on computer architecture was presented in (Schagaev et al. 1989) and still in full swing in use in good universities around the globe for master and Ph.D. research.

Since then, nothing much has been changed in architecture, few system architectures were lost in market competition (not the worst one, to be honest), leaving SPARC, Intel, ARM and SPARC with lion shares of the market.

The SPARC is included in this comparison, as it is heavily used in space, especially the fault-tolerant version LEON3 (LEON3 2009).

To have a consistent comparison, we only compare the 32-bit versions of the respective processors, although 64-bit versions are available for the × 86 and SPARC architecture.

Table 6.1 Available instructions

Instr name	Format code	Inst. code	Op1	Op2	Description
NOP	01	0000	Ignored	Ignored	Execute no action except increasing PC
STOP	00	0000	0	0	Stop instruction execution
TRACE	00	0000	$R_i > 0$	$R_j > 0$	Output R_i to debugger Operand 1 or Operand 2 must be > 0
RETI	11	0000			Return from interrupt
LD	11	0001	R_i	R_j	Load 32-bit memory at address R_i into register R_j ($R_j := *R_i$)
LDA	11	0010	Ignored	R_j	Load the value from the next 32-bit word ref to PC and store it in R_j ($R_j :=$ constant) Operand 1 is ignored
ST	11	0011	R_i	R_j	Store content of register R_i to the memory at address R_j ($*R_j : =$ constant). Operand 1 is ignored
MOV	XX	0100	R_i	R_j	Move content of *register* R_i to *register* R_j ($R_j := R_i$)
ADD	XX	0101	R_i	R_j	Arithmetically add the content of R_j to the content of R_j and store the result in $R_j := Rj + Ri$
SUB	XX	0110	R_i	R_j	Arithmetically subtract the content of R_j to the content of R_j and store the result in $R_j := R_j - R_i$
ASR	XX	0111	R_i	R_j	Shift the content arithmetically one bit right and store in R_j
ASL	XX	1000	R_i	R_j	Shift the content arithmetically one bit left and store in R_j
OR	XX	1001	R_i	R_j	Perform a bit wise logical OR of register R_i with register R_j and store the result in R_j
AND	XX	1010	R_i	R_j	Perform a bit wise logical AND of register Ri with register *Rj* and store the result in *Rj*
XOR	XX	1011	R_i	R_j	Perform a bit wise logical XOR of register R_i with register R_j and store the result in R_j
LSL	XX	1100	R_i	R_j	Shift the content of register R_j logically one bit to the left and store the result in R_j
LSR	XX	1101	R_i	R_j	Shift the content of register R_j logically one bit to the right and store the result in R_j
CND	XX	1110	R_i	R_j	Arithmetic comparison of R_i with R_j and store the result in R_j
CBR	XX	1111	R_i	R_j	Jump to address in R_j if R_i is nonzero and save PC in R_i

Table 6.2 Check of conditions

Bit	0	$R_1 > R_2$
Bit	1	$R_1 > R_2$
Bit	2	$R_1 = R_2$

Table 6.3 Relation of operands as a result of comparison

Relation	Bit mask
$<$	010
$\leq$	110
$>$	001
$\geq$	101
$=$	100
$\neq$	011

Both the SPARC and the ARM processors are RISC-based and are thus quite similar in the instruction set, whereas the × 86 architecture is CISC-based and provides a multitude of instructions in comparison to the other architectures.

The × 86 is also the only architecture that allows using memory locations directly in instructions (register-memory architecture), whereas RISC machines (load and store architecture) must first load the argument into a register.

This enormous number of instructions leads to the curious situation that the instruction decoder of the Intel Atom CPU needs as much space as a complete ARM Cortex-A5.

6.4.1 Processor Architectures: The Overview

Table 6.4 gives an architectural overview of the mentioned architectures. The overview and analysis are inspired and adapted from the architectural comparison in (Schagaev et al. 1989). All information to compile the table was acquired from various documents from Intel, ARM, LEON documentation mentioned already, as well as two monographs of Hennesy and Patterson.

The processor architecture summary is given in Table 6.4 and clearly illustrates the simplicity of the ERA processor architecture with its 16 instructions, which is by a large margin smaller than the other architectures.

However, the simplicity does not go free. It comes at the cost of less powerful instructions in comparison to the other instruction sets and resulting thus in longer code. With longer code, we mean that the number of instructions is higher, not necessarily the code size, as the instructions are only 16-bit wide.

The load from memory serves as a typical example where in case of array accesses, the absolute memory address must be explicitly calculated and stored in a register prior to the memory access, where, for example in the SPARC architecture, the offset

Table 6.4 Comparison of instruction sets

	ERA	×86	SPARC v8	ARM7TDMI (ARMv5-TE)	ARM7TDMI (thumb)
Date announced	2006	1978	1987	1985	1995
Architecture family	Reduced RISC	CISC	RISC	RISC	RISC
Integer registers	32 × 32 bits	8 × 32 bits	31 × 32 15 × 32 bits	15 × 32 bits	8 × 32 bits + SR, LR
Floating-point registers	0	Optional 8 × 32 bits or 8 × 64 bits (8 × 80 bits internal)	32 × 32bits or 16 × 64 bits or 8 × 128 bits	Optional 32 × 32 bits or 16 × 64 bits	Optional 32 × 32 bits or 16 × 64 bits
Vector registers	0	Optional 8 × 64 bits or 8 × 128 bits	0	Optional 32 × 32 bits or 16 × 64 bits	Optional 32 × 32 bits or 16 × 64 bits
Address space	32 bits, flat	32 bits, flat or segmented	32 bits, flat	32 bits, flat	32 bits, flat
Instruction size (bytes)	2	1–15	44	4	2
Multiprocessor capable	No	Yes	Yes	Yes	Yes
Memorymanagement unit	No	Yes	Optional	Optional	Optional
Memoryaddressing mode	1	7	2 (reg ind_ + disp, reg ind indexed)	6	6
Memoryaddressing sizes	32	8, 16, 32 bits	8, 16, 32, 64 bits	8, 16, 32, 64 bits	8, 16, 32, 64 bits
Number of instructions	16	138 integer and logic 92 floating point, total 332	72	53	37
I/O	Memory mapped	Memory mapped	Memory mapped	Memory mapped	Memory mapped

(continued)

Table 6.4 (continued)

	ERA	×86	SPARC v8	ARM7TDMI (ARMv5-TE)	ARM7TDMI (thumb)
Pipeline length	No pipeline	Atom: 16, Corei7/ Core2: 14, Pentium 4; 20–31	Leon 3: 7 SPARC 64V-15 SPARC T2-8 Registers	3	3
Specialities	Very simple instructions, built-in fault tolerance	Instructions of variable size, grown instructions set, memory operands	Register window delayed control transfer	Conditional instruction execution	32-bit ARM instructions partly required; solved with Thumb up mode

to the base address can be stored in a distinct register and then added on the fly in the load instruction itself. The ARM even allows encoding an offset to a base address given in a register directly in the instruction itself.

The ARM Thumb instruction set is a subset of the standard 32-bit ARM instruction set. It is intended for resource constraint environments where only a 16-bit data bus is available, and it increases the code density. The Thumb mode does not change the ARM architecture at all, in specific the address space is still 32-bit and all registers are 32-bit wide.

However, only registers R0 and R7 are directly accessible, and R8 are R16 are hidden. The stack register (SR) and the link register (LR) are accessed by some instructions.

Interestingly, exceptions (interrupts) always trigger a processor switch to the 32-bit ARM mode prior to executing the exception handler.

This supports the statement by Hennessy and Patterson (1996, 2002) that the 16-bit mode is not supposed to be a full architecture, but "they [the instructions] are enough to encode most procedures".

In contrast to this, we say that the ERA instruction set, although it is even more constraint than known RISC architectures.

Although some, such as SPARC, do not directly provide an absolute addressing mode, the register + register mode with a nullified second register is used to emulate this. The SPARC architecture even provides a register which is always nullified; thus, the absolute addressing does not incur any performance penalty.

The ×86 addressing modes are more powerful than the RISC ones; thus, instead of extending the above table, the ×86 addressing modes are summarized in Table 6.5.

The offset part of a memory address can be either a static displacement or through an address computation made up of one or more elements. The resulting offset is called effective address and can constitute of either positive or negative values except for the scaling factor. The ERA architecture provides only direct memory addressing, which results in longer code as the addresses must be explicitly computed before the data can be loaded.

Table 6.6 compares the offset sizes that are directly encoded in the instruction. If not mentioned differently, the offset is always relative to the current instruction pointer.

Table 6.5 Supported addressing modes

Data addressing mode	ERA	× 86	SPARC v8	ARM	ARM Thumb
Register	X	X	(X)	X	X
Register + offset (displacement of based)	–	X	X	X	X
Register + register (indexed)	–	X	X	X	X
Register + scaled register (scaled)	–	X	–	X	–
Register + offset and update register	–	–	–	X	–
Register + register and update register	–	–	–	X	–

Table 6.6 Offset size of encoded instructions

Offset size encoded in instruction (in bits)	ERA	× 86	SPARC v8	ARM	ARM thumb
Unconditional jump/ call	0	8–32 signed, rel. or absolute, direct or indirect	30	24	11
Conditional branch	0	8–32 signed	19	24	8

The ERA architecture does not allow encoding the offset in an instruction, but must always be given as an absolute address in a register. Thus, it requires an additional "load constant" instruction, which requires another 8 bytes (Load + NOP + 4 Byte constant). In terms of code safety, absolute jumps are preferable as calculated jumps (relative) are more susceptible to faults than absolute jumps.

Thumb instruction set, is intended to be used as a full instruction set, generic enough to encode all language features.

Compared to the ARM, the ERA has less than half the number of instructions, making it much simpler.

On the downside, more elaborate instructions must be emulated, such as relative memory accesses or procedure calls. In particular, the latter is implemented on the ARM by the "load multiple" and "store multiple" instructions that allow for efficiently putting all procedure arguments on the stack. ERA, on the other hand, must store all arguments individually on the stack.

Although some, such as SPARC, do not directly provide an absolute addressing mode, the register + register mode with a nullified second register is used to emulate this. The SPARC architecture even provides a register which is always nullified; thus, the absolute addressing does not incur any performance penalty.

The × 86 addressing modes are more powerful than the RISC ones; thus, instead of extending the above table, the × 86 addressing modes are summarized in Fig. 6.9.

The offset part of a memory address can be either a static displacement or through an address computation made up of one or more elements. The resulting offset is

$$offset = \begin{Bmatrix} CS: \\ DS: \\ SS: \\ ES: \\ FS: \\ GS: \end{Bmatrix} \begin{bmatrix} EAX \\ EBX \\ ECX \\ EDX \\ ESP \\ EBP \\ ESI \\ EDI \end{bmatrix} + \left[\begin{pmatrix} EAX \\ EBX \\ ECX \\ EDX \\ EBP \\ ESI \\ EDI \end{pmatrix} * \begin{pmatrix} 1 \\ 2 \\ 4 \\ 8 \end{pmatrix} \right] + \begin{bmatrix} None \\ 8-bit \\ 16-bit \\ 32-bit \end{bmatrix}$$

$$offset = Selector : Base + (Index * Scale) + Displacement$$

Fig. 6.9 × 86 addressing modes

called effective address and can constitute of either positive or negative values except for the scaling factor.

The ERA architecture provides only direct memory addressing, which results in longer code as the addresses must be explicitly computed before the data can be loaded. The ERA architecture does not allow encoding the offset in an instruction but must always be given as an absolute address in a register. Thus, it requires an additional "load constant" instruction, which requires another 8 bytes (Load + NOP + 4 Byte constant).

In terms of code safety, absolute jumps are preferable as calculated jumps (relative) are more susceptible to faults than absolute jumps.

Table 6.7 compares the instructions required in each architecture to perform basic RISC operations such as load, stores.

A sequence of instructions is given if an instruction is not available in the instruction set. A "-" means that more than a short instruction sequence is required to simulate the specific functionality, typically a longer sequence of inline code or a library call. As the ERA architecture does not include a floating-point unit, we omit here all floating-point instructions, as all of them are emulated in software.

6.4.2 Passive Zone

Figure 6.10 presents the current custom hardware prototype of the ERA device. The prototype is provided with two flash-based ROM modules 128 Mb each (8 Mb × 16 bit), with replicated bootstrapping firmware and operating system software.

The proposed memory scheme may be regarded as a collection of 4 blocks, 16-bit wide, with identical size of 1 Mb each (16 × 64 k). Using 16-bit memory modules instead of 32-bit memory modules increases reliability and reduces, when necessary, energy required for execution. Additionally, the scheme includes two flash-based ROM modules with replicated bootstrapping firmware and operating system software.

Reliability is increased by means of added working states and configurations.

Energy-wise operation is improving by means of this architecture ability to activate only modules required—by means of using a single 16-bit memory module, when necessary.

The proposed memory scheme allows different configuration schemes, explained in 6.4.3.

6.4.3 Interfacing Zone

The above-explained architecture principles of design relate to the interfacing zone as well. The architecture needs something (logic) in the interfacing area to enhance

Table 6.7 Comparison of selected instructions

Instruction	ERA	× 86	SPARC v8	ARM7TDMI (ARMv5-TE)	ARM7TDMI (thumb)
Load word	LD	MOV	LD	LDR	LDR
Load byte signed	–	MOVSX	LD	LDRSB	LDRSB
Load byte unsigned	LD, LDA, AND; –[a]	MOV	LDUB	LDRB	LDRB
Store word	ST	MOV	ST	STR	STR
Store byte	– b	MOV	STB	STRB	STRB
Add	ADD	ADD	ADD	ADD	ADD
Add (tap if overflow)	–	ADD, INTO	ADDcc, TVS	ADDS, SWIVS	ADD, BVC + 4 SWI
Sub	SUB	SUB	SUB	SUB	SUB
Sub (trap if overflow)	–	SUB, INTO	SUBcc, TVS	SUBs SWIVS	7SUB, BVC + 4, SWI
Multiply	–	MUL, IMUL	MULX	MUL	MUL
Divide	–	DIV; IDIV	DIVX	–	–
AND	AND	AND	AND	AND	AND
OR	OR	OR	OR	ORR	ORR
XOR	XOR	XOR	XOR	EOR	EOR
NOT	–	NOT			
Shift logical left	SLL[c]	SHL	SLL	LSL	LSL
Shift logical right	SLR[c]	SHR	SRL	LSR	LSR
Shift arithmetic right	SAR[c]	SAR	SRA		
Compare	CND	CMP	SUBcc	CMP	CMP
Conditional Branch	CBR	JG; JNLE; JGE; JNL; JL; JNG; JNO;JNS; JO;JS	BR..Z, BPcc (<, >, ≥, ≤, =, *I* =)	B	B
Call	CBR	CALL	CALL	BL	BL
Trap	CBR	INT n	TIcc, SIR	SWI	SWI

(continued)

Table 6.7 (continued)

Instruction	ERA	× 86	SPARC v8	ARM7TDMI (ARMv5-TE)	ARM7TDMI (thumb)
Return from Interrupt	RETI	IRET	DONE; RETRY, RETURN	MOVS Pc, r14	−d
NOP	NOP	NOP	SETHI r0, 0	MOV r0,r0	MOV r0,r0

[a] If the 8-bit data is aligned, the sequence LD, LDA, AND can be used, otherwise, a LD and appropriate number of shift operators must be used
[b] For storing an 8-bit value, the destination address must be loaded, the appropriate bits cleared by using a bit mask, the arguments shifted and then written back to memory. In other words, omit using 8-bit values
[c] Only one bit
[d] Interrupts are always handled in 32-bit mode. Thus, a pure Thumb CPU without 32-bit mode is not supported

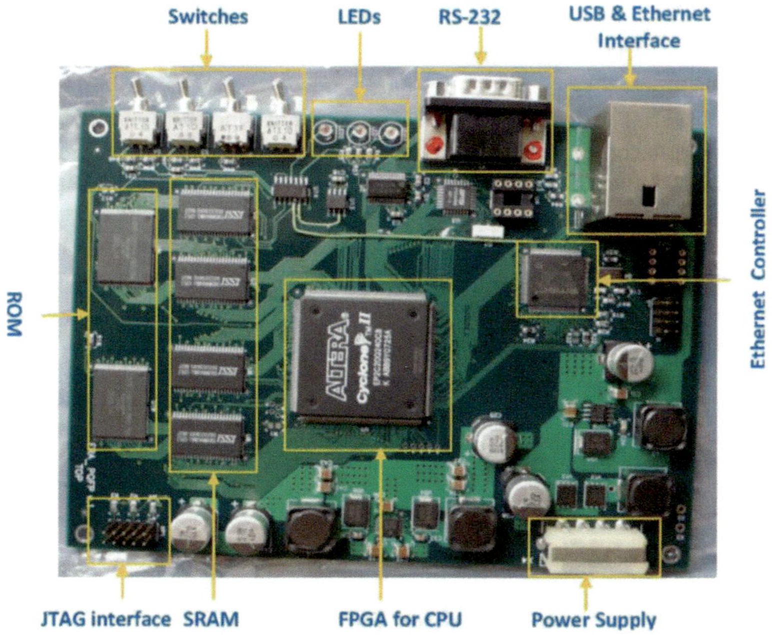

Fig. 6.10 ERA prototype board

the flexibility and resilience of system operation between active and passive zones, for a wide range of applications where PRE-properties are key requirements.

This motivates the ability to have a reconfigurable interfacing zone. One of the schemes that we propose is known as T-logic. T-logic is a hardware element that provides reconfigurability of the architecture for performance, reliability and energy-wise operation. This logic should be able to provide minimal configuration—when one processor and memory remain.

Due to importance of T-logic as "major executive agent" for reconfigurability of the whole system, the next section explains it in detail.

6.5 ERA Reconfigurability

As declared earlier: performance, reliability and energy-awareness are required for the next generation of computer systems. Reconfigurability, as a new feature, requires hardware and system software implementation and support. In order to be able to change the configuration when necessary (sometimes several times during a single mission) systems should have special elements with specific properties such as extreme reliability, performance and simplicity, supported by independence from faults of the system.

For the purposes declared above we propose a hardware element called T-logic. The main function of a T-element is to connect and disconnect the system component, using "logic rotation" for various types of connections and configurations.

In order to connect and disconnect the system component there are several "turns" of T-element as Table 6.8 shows, describing the basics of T-Logic element.

The reconfigurable interconnection schemes can execute dynamic reconfiguration of system transparently from for software. For example, when we use triple system elements configuration, T-elements might exclude faulty ones from operation, leaving only 2 elements active (DMR). If further degradation is allowed, the configuration (in terms of fault tolerance) can continue up to a single element (Schagaev and Buhanova 2001).

Figure 6.11 illustrates how reconfiguration might be used for reliability purposes using T-elements. The figure shows a diagram that exemplifies a scheme with 3 memory modules working in parallel that at some point during operation, experiences two permanent failures in two out of three modules.

6.5.1 *T-Logic for Memory Management*

A power consumption of ERA can also be controlled using T-elements. Existing electronic technologies possess the following drawback: growth of power consumption causes degradation of system reliability. Therefore, an ability to connect and disconnect a system element might be function and requirement in real time and other applications.

Table 6.8 T-logic rotation

Position	Description
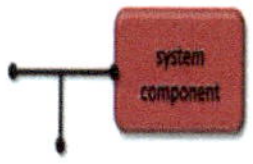	The "T" logic connects the active zone to a front element that is connected in redundant mode with a right element This front element leads the rest of the elements that is connected to
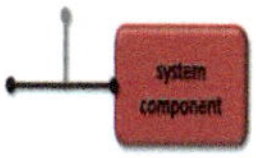	The "T" logic connects the element to two neighbouring elements in redundant mode. The element is leaded by the left side component that is connected to the active zone
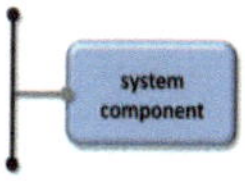	The "T" logic connects the element to a left element system component will be leaded by side element that is connected to the active zone
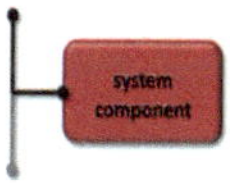	The "T" logic connects the element to a left element system component will be led by side element that is connected to the active zone
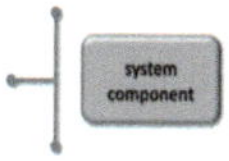	Disconnected element: The "T" logic is disconnected from the interconnection scheme. The energy consumption of the element is reduced to the minimum

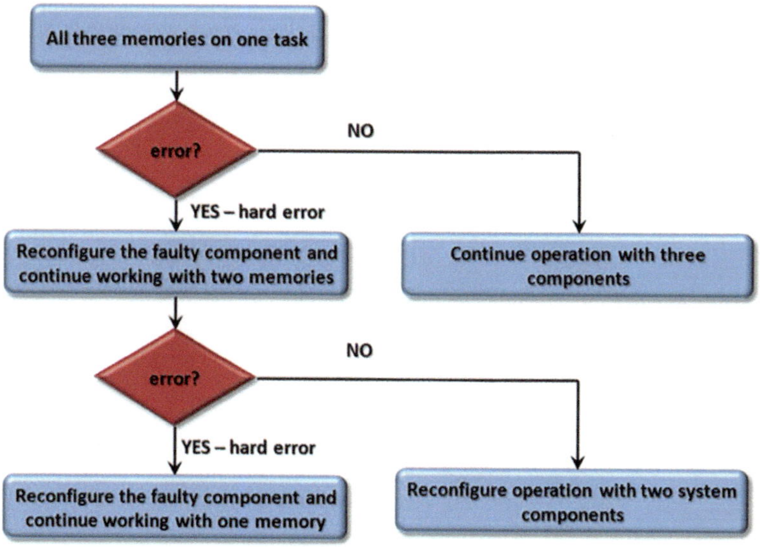

Fig. 6.11 Algorithm of configuration for reliability using T-logic

Again, for illustrative purposes, we use an example when ERA uses three system elements in redundant mode. If the task scheduled does not require a full size configuration, the architecture can be configured to operate using either two elements or a single element on its own.

Hardware configuration should be implemented transparently to application programs. Task defined requirements might be available for a run-time system. Excluding memories will not change the logic of the program.

Figure 6.11 also illustrate that the power saving algorithm for a reconfigurable system is possible. The reconfigurable interconnection schemes improve the memory management flexibility. For instance, if the computer architecture has three elements and the task requires maximum capacity, the system will configure all memories in redundant mode so that it can provide maximum capacity.

6.5.2 T-Logic for Configuration in ERA

Possible configurations of a system that uses "T" logic are presented in Table 6.9.

Table 6.9 Possible system configurations using T-logic

Configuration	Explanation
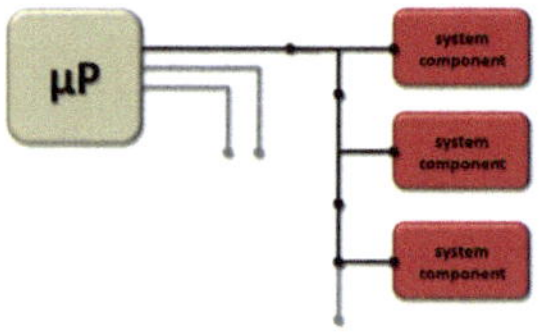	"T" configurators connect all three components with processor. Top system component acts as leading element. The rest system elements compare the results and participate in voting. Thus reliability of this system configuration is high. "T" element configures enables excluding faulty system component
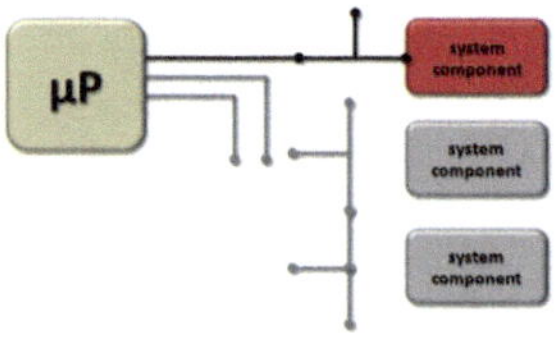	This system configuration serves for maximum energy saving. In this case "T" element connects only one system component with processor, while the rest are idle
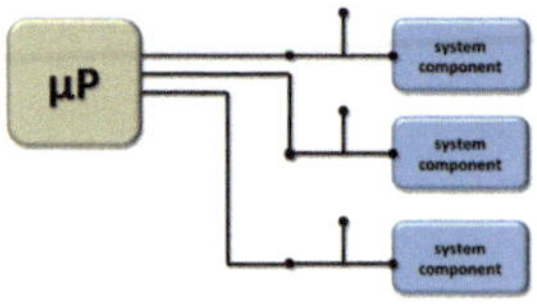	In this case, all three components are used for maximum hardware capacity. When performance of application is main priority, this configuration fits the purpose

The first Raw illustrates maximum reliability configuration (if 3 HW elements are available). The power saving of the system could be improved by disconnecting elements from system and keep them idle—raw 2.

The configuration for maximum capacity required for application is shown in the third raw.

To provide fail-safe functioning for ERA T-logic itself, it should provide the following property:

- When "T" element itself fails system functioning should continue. Its design, therefore, must implement fail-safe principle; in this case T logic element should be replaced by wire to enable further system functioning. This all, in fact defines requirements for interconnection zone for ERA PRE-smart functioning.

This illustrates how *principle of separation of concerns* can be applied for design of hardware schemes.

6.6 Syndrome

As mentioned earlier, the system new property must be supported by hardware and system software implementation of required processes that make this property.

We introduce for this purpose a special hardware scheme called *a syndrome*. The term *Syndrome* is a new Latin (origin 1535–45) and originated from Greek "syndrome" where:

1. "Syn-" from combination, concurrence
2. "Dramein", main meaning is "to run"

For our purposes, a syndrome is not passive, presenting "a snapshot status" of a system, a *Syndrome* is also active, a serving tool to control the system configuration.

Thus, a *syndrome* is "*a group of related or coincident things, events, actions, signs and symptoms that characterize a particular abnormal condition*".

Further analysis and development of syndrome concept and application follows.

6.6.1 Syndrome Use

Clear that functions of a syndrome are not only passive, presenting "snapshot-status", but also active, serving as a tool to control of the system configuration and enabling to estimate system conditions.

Syndrome also might help to answer a question that is shamefully omitted in the vast majority of research about fault tolerance:

WHAT PROVIDES THE FAULT TOLERANCE OF THE SYSTEM? ! ! Q1

It is usually assumed that the core logic is ultra reliable and guarantees control of configuration and reconfiguration. Unfortunately, using the homogeneous redundancy may limit the increase in reliability since techniques based on the same type of redundancy are vulnerable to the same threats.

Hybrid techniques based on heterogeneous redundancy can be more effective.

Thus, even when checking schemes of memory or configurators or processor detect error and transfer information about it to the syndrome this information might not be useful if the system does not include either one or both mentioned below:

1. "External elements" responsible for exercising GAFT and making decision on configuration/reconfiguration if necessary, or
2. Internal element that capable initiate required sequence.

Indeed, in regular computing system when there are faults in the processing logic, to expect to expect that it is able to perform self-healing and then control and monitor configuration of the rest of the system is an attribute of fairy tale, not engineering and seems to be unrealistic. There is a solution though, as described below.

Figure 6.12 shows conceptually ERA active zone divided by two AU and LU elements.

To be able to absorb and trust an information about a status of an element, every checking signals about condition of registers (not shown) memory, AU and LU, as well as control unit should be aggregated in syndrome.

The scheme of implementation of fault tolerance separates the passive zone and active zone of the proposed architecture. A clear separation of the functions of processing (of data operation) and storing (memory) enables to apply various checking and recovery solutions. The passive zone has the elements to store data, while active zone is for data manipulation.

All processor registers (register file) may be protected by parity or other checking schemes. During instruction execution data are loaded from the register file into functional elements for operation and operands are checked.

If there is no fault detected during instruction execution the operation is considered as successful and the result is stored back into the register file or sent out to memory. The fast and reliable memory access is possible by application of static memory and register files.

The memory context data might be protected by schemes such as error checking code, Hamming code, or recently proposed schemes (Gössel et al. 2008).

If error is detected, the control unit by executing GAFT attempts to restore the damaged data and repeat the instructions:

1. Restore the damaged data
2. Repeat the execution of the instruction that manifested a fault
3. Resumed execution.

ERA provides the fast and reliable recovery scheme within instruction execution level, transparently for software.

Taking into account requirement of minimization of redundancy, the architecture presented in Fig. 6.12 seems to be efficient—there is no duplication or greater level of

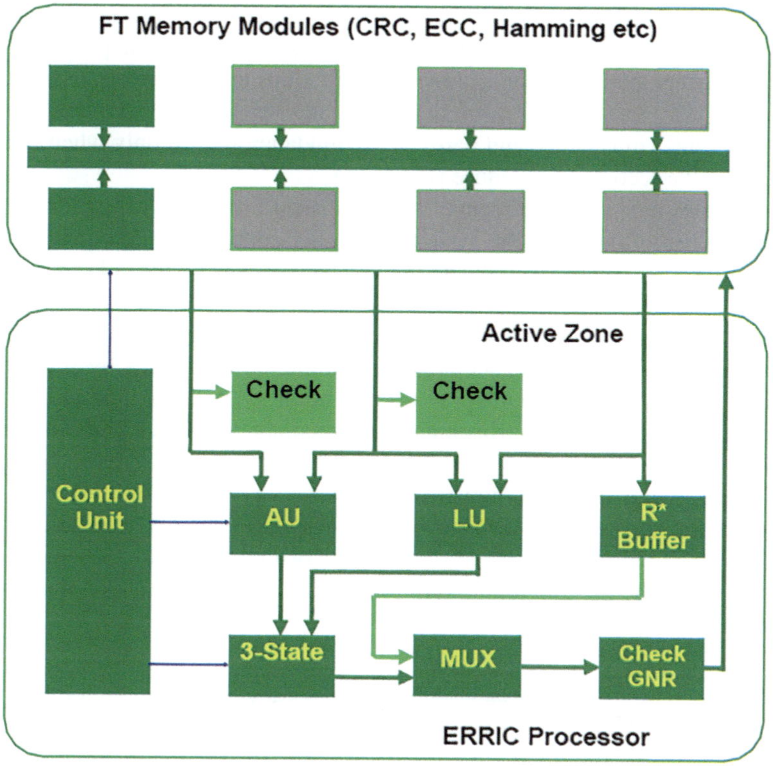

Fig. 6.12 Processor-memory structures with "separation of concern"

reservation applied for active zone. Active zone consists of two non-identical units: arithmetic and logic units, respectively, denote them (AU and LU).

In terms of power consumption this scheme is also efficient. A question 2—Q2 arises:

HOW TO TOLERATE FAULTS OF AU AND LU WITHOUT DUPLICATION? Q2

The question Q1 above becomes crucial for implementation of fault tolerance for any system and proposed scheme with minimum redundancy.

The solution is possible if we apply well-known mathematical results about equivalence of arithmetic polynomial for logic functions (Malyugin 1982), attempting to represent arithmetic functions by sequence of Boolean functions.

There is a theory when arithmetic functions are represented by Boolean functions. These two groups of results hint us a possible solution:

An ISA that consists of logic and arithmetic instructions should be supported by sequence of functional equivalence of logic operators implemented by arithmetic unit. For every arithmetic instruction a functionally equivalent sequence of logic instructions should be added.

Both sequences should be stored in different segments of read-only memory. If a fault occur in the AU a signal sets a flag in the syndrome and sequence of logic operators is executed instead of arithmetic instructions to complete GAFT, and vice versa.

Note that this equivalence and hardware redundancy is used only when a fault is detected and the recovery procedure is initiated.

This enables the system to recover from transient errors and execute reconfiguration. After reconfiguration, the system can either continue functioning in normal mode or, in case of an unrecoverable permanent fault, it can provide fail stop sequence of actions.

Area of applications for embedded safety critical systems set hard constrains for instructions execution and replacement of arithmetic instructions by sequence of logic ones has no justification to apply for execution of application programs.

The syndrome acts as a control centre for three main functions: fault monitoring, reconfigurability and recovery (Fig. 6.13). These three functions serve for the purpose of performance, reliability and power efficiency.

From a hardware point of view, the syndrome is represented as a special hardware register that will interact with the system via hardware interruption schemes.

Semantically, the structure of the syndrome is subdivided into three different areas (Fig. 6.14):

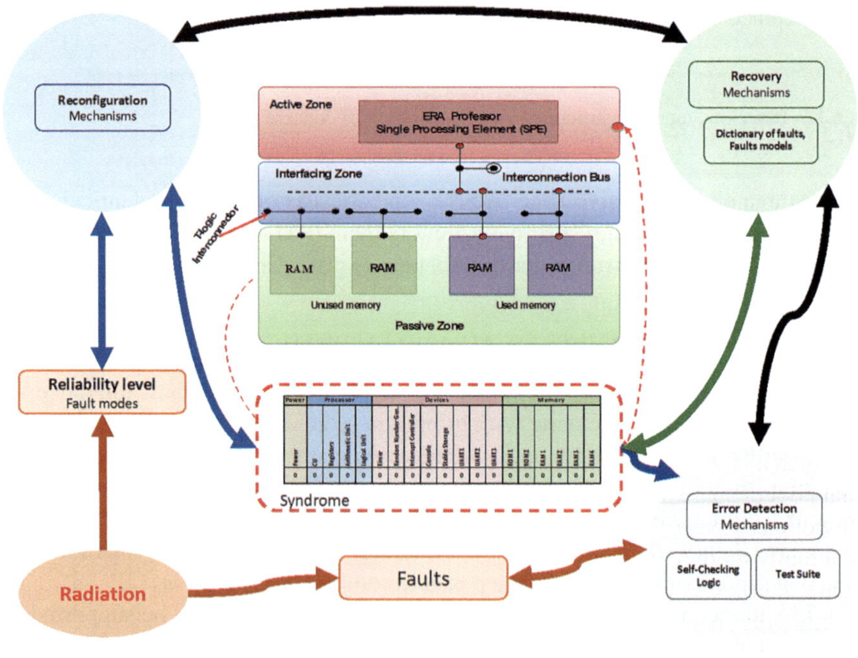

Fig. 6.13 Syndrome purposes

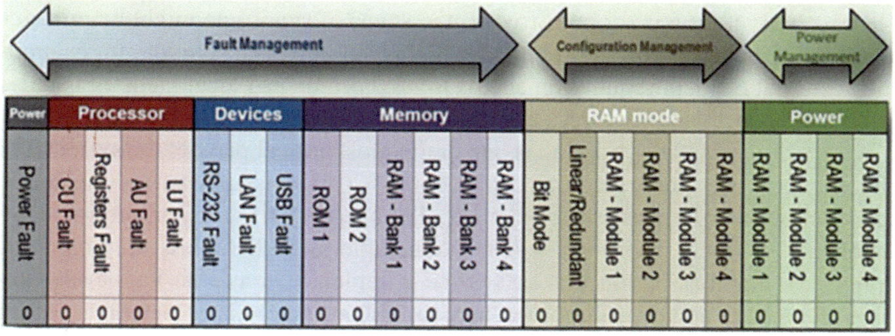

Fig. 6.14 Syndrome for reconfigurable architecture

1. Fault control area
2. Configuration control area and
3. Power control area.

The Fault control area reflects the hardware status of the different areas of a single processing element: processor, memory and interface. Full "Zero" syndrome in this area indicates that no fault has been detected in the system. If a fault in a specific element is detected, the corresponding bit is set to 1.

Each hardware element has its own representative flag within syndrome. For ROM memory group that consists of two chips syndrome of ROM condition has two positions with zero when ROM works correctly. In turn, static RAM memory has 4 hardware elements.

The configuration area of syndrome reflects memory mode that ERA is currently using. The Bit mode field defines whether the addressing mode is 16 or 32 bits, whereas the L/R field defines whether the memory banks are in linear or redundant mode.

Bit Mode: 0 = 16 bits
1 = 32 bits addressing mode
LR: 0 = Linear
1 = Redundant.

The power management area reflects the status of the modules in terms of power.

Bits Power RAM Module 1, 2, 3 and 4 represent whether the Memory module is powered:

0 = Power Off
1 = Power On.

The combination of those three areas: Fault, configuration and Power management control defines the state of the system. For example, a memory module could be in the following states: faulty, failed, stand-by, ideal and off state. Checking those 3 areas by a simple reading would determine the status of the memory module.

When reconfiguration is set by software, the states of syndrome might be mirrored in memory. Keeping those states only in SSW is not universal solution, for example when external element requires an access to a syndrome and local memory is failed further use of available resources and elements becomes impossible.

Without a doubt, the syndrome is one of the most critical parts of the system. For reliability purposes, there are three copies of the 32-bit register syndrome connected to a voter within the processing element. Triplication of the syndrome increases complexity of logic. The voter will be vulnerable as well (unless the voters are triplicated). Another option that solves the complexity would be low-level hardening techniques and/or using different technologies (such flash memory) just for the syndrome register.

However, that will increase the manufacturing costs. Without this replication, and possibly different, enhanced technology use a bit flip in the fault area of the syndrome would lead to redundant fault detection processing, whereas a bit flip in the configuration area would likely end up in a catastrophic failure.

6.6.2 *Location Access and Way of Operation of the Syndrome*

There are two major mechanisms that will be able to detect a fault: hardware logic and SSW. Both, hardware logic (mismatch and voters) directly and SSW (testing and detection mechanisms) via instruction should have access to the syndrome:

SSW events: If SSW has access to the syndrome, reading and writing the value of the syndrome are compulsory. Note that the current ISA does not include instructions within the processor to do that, unless one of the registers (e.g.: register 31) is used as a syndrome.

This is similar to the VAX V70 processor that includes a status word placed in a fixed address of the memory. In this case, the register file needs to be hardened. At least the syndrome needs to be hardened.

A bit flip in the configuration area of the syndrome could mistakenly turn off one of the memory modules.

As a consequence, a preferred way to access the syndrome that avoids changing the ISA and hardening the register file is the use of the input/output memory lines (mapped in a reserved address). This works as a TMR scheme, with complete software independence.

Regardless of the hardware implementation, only one syndrome is visible to the system software. An error in one of the syndrome registers is corrected by the hardware without the software intervention.

However, it is useful that the SSW is aware of errors in the syndrome. Errors within the syndromes are useful information in a potential contingency plan (e.g.: setting the fault tolerance of the system to a higher level in case of recent particle impacts).

Automatic events via hardware detection mechanisms. Special HW interruptions are needed for this; if during the diagnosis of a memory chip the ALU, for example signals a problem, a diagnosis of the suspected element should be done first.

The syndrome might be considered as independent new hardware—independent from processor hardware elements and other schemes. A method is needed for synchronizing the operation of the processor with the syndrome. One solution would be polling, where a loop that checks the status of the syndrome is arranged. It has a major disadvantage: the processor is busy reading the syndrome, instead of executing some useful code (wasteful in terms of processing power).

Instead of polling the syndrome waiting for a change, a hardware interruption system is preferred. In this case, the syndrome subsystem is responsible for notifying its current state to the processor. When the syndrome needs the processor's attention, it sends an electrical signal through a dedicated pin in the interrupt controller. In this case, the processor stops its current activity and jumps to execute a function (interrupt handler) that has to be associated with the fault manifestation.

By using hardware interruptions, in terms of total execution time, the syndrome will be accessed only when a fault is manifested and reflected in the syndrome. Most of the time, the fault area of the syndrome will be 0 and the rest of the areas will only be altered when restarting the system or when changing the memory mode.

The same method might be applied to the control of other devices. Different levels of interruptions are then needed. Active zone hardware should have higher priority than passive zone hardware.

If a hardware mechanism sets the syndrome bits, then the processor executes a trap (exception) and starts the diagnosis software. The software treats the fault and clears the syndrome otherwise the processor would trap again and restart the diagnosis.

This scheme introduces requirements to set and re-set syndrome register internally (using own hardware or software) or externally, from the "rest of the system" when ERA is used in the form of CC—connected computer structure.

If the syndrome is located within the active area, in the case of a faulty active area a neighbour processing element may have difficulties accessing to it. To resolve it there are three options:

1. To enable system flexibility in fault handling, one has to implement syndrome independently from the active area where it can be accessed via hardware by neighbour single elements.
2. Software message passing: Instead of hardware access SSW will deal with status of single elements, by sending un update (periodic updates) on syndrome to the single element neighbours before changing memory mode or when a fault has been detected (if feasible in this last case). A syndrome table in a similar fashion (but not as heavy) to the routing tables used by routers and the different routing algorithms could be shared by different processing elements.
3. Both, Hardware access and SW message passing are not mutually exclusive. A combination of both is possible.

6.6.3 Syndrome: Passive Zone Configurations

A total of 25 operating states of memory are possible to operate in reliability, energy or performance wise modes. The characteristics of the proposed memory architecture are given in Tables 6.5 and 6.10.

Memory modes are subdivided into two major categories, depending on the number of bits read/written at once: 32 bits and 16 bits. At the moment there are 10 different usable configurations in 32-bit mode memory (defined in Table 6.6) and 15 different configurations in 16-bit mode memories (defined in Table 6.5).

In addition, two modes of operation, depending on the existing amount of redundancy, can be selected:

1. Linear mode: where no module is replicated and
2. Redundant mode: with at least one module being replicated.

In order to explain the available memory modes, let's consider the letters in the set {A, B, C, D, x} as a representation of information stored in the 16-bit memory modules, where:

1. A lowercase letter x represents a module is not in use.
2. Identical letters represent identical information in different modules. In other words, information stored in a module is n times replicated into n modules: e.g. AA, AAA, BB and AAAA. In this case $n - 1$ modules are connected in shadow mode and perform all memory operations concurrently to their respective master memory module. A hardware voter in the memory controller compares the output of the memory modules and triggers a fault in the syndrome in case of mismatch.
3. The pairs AB, BA and CD represent two 16-bit modules combined into a virtual 32-bit module.

6.6.3.1 32-Bit Mode

The memory modules of ERA are 16-bit wide. Therefore, two memory chips must be combined to allow 32-bit memory access. Table 6.10 reflects all the supported memory combinations in 32-bit mode. Only one memory mode is available in 32-bit Linear Mode (state 10 in Table 6.6). Module 1 & 2 are combined and mapped in the memory space. Module 3 & 4 are combined as well and mapped contiguously in the memory. This configuration provides maximum space for the application but does not feature fault tolerance, and not even fault detection at the memory level.

In terms of fault tolerance, the maximum redundancy based on 32-bit addressing is the ABAB configuration (Table 6.11). The information in the two modules AB is replicated on an extra pair AB. In every reading operation the data from both memory pairs is compared. If there is a mismatch, the checkpoint area in RAM will be the decisive factor in selecting the failure.

Note, that 32-bit modes that involve 3 working modules, such as ABAx ABxB, AxAB or xBAB are not initial modes (starting in this mode does not offer any

Table 6.10 16-bit addressing modes in RA

State	Phase	RAM mode		Information-power								
		Bit Mode	Linear/ Redundant	RAM-Module 1	RAM-Module 2	RAM-Module 3	RAM-Module 4	Space (Mb)	Reliability	Speed	Power consumption	Syndrome
11	4	16	L	A				1	1	2.5	2	111000
12	4	16	L		A			1	1	2.5	2	110100
13	4	16	L			A		1	1	2.5	2	110010
14	4	16	L				A	1	1	2.5	2	110001

Table 6.11 A 32-bit addressing modes in RA

State	Phase	RAM Mode		Information-power				Space (MB)	Reliability	Speed	Power consumption	Syndrome
		Bit Mode	Linear/ Redundant	RAM-Module 1	RAM Module 2	RAM Module 3	RAM Module 4					
1	1	32	R	A	B	A	B	2	2	4	4	111111
10	3	32	L	A	B	C	D	4	1	5	4	101111
2	2	32	R	A	B	A		2	1.5	4	3	111110
	2	32	R	A	B		B	2	1.5	4	3	
	2	32	R	A		A	B	2	1.5	4	3	111011
	3	32	R		B	A	B	2	1.5	4	3	110111
	3	32	R	A	B			2	1	5	2	111100
	3	32	R			A	B	2	1	5	2	1110011
	3	32	R	A			B	2	1	5	2	111001
			R		B	A		2	1	5	2	110110

advantage). Those four modes would typically involve that an error has occurred in one of the modules ongoing diagnostics are taking place.

6.6.3.2 16-Bit Mode

With the main purpose of critical energy saving, a 16-bit addressing mode was introduced. This mode uses a single bank of memory. Hence, there is not duplication of memory. The four different memory configurations, currently allowed in this mode are presented in Table 6.11.

When only one single mode is available due to permanent faults in the other three, the system could be restarted in a fresh 16-bit mode loading the different SSW binary codes from a ROM location into RAM. Note that this mode is an emergency mode and does not contemplate the possibility of hot switching from a 32-bit mode. Schemes to change from 32-bit mode to 16-bit/power saving mode and vice versa are defined by the runtime system.

6.7 Graceful Degradation

If one of the memory modules fails, the system can be reconfigured to exclude it from the current configuration. In terms of fault tolerance, the maximum redundancy based on 32-bit addressing is the ABAB configuration (Fig. 6.16).

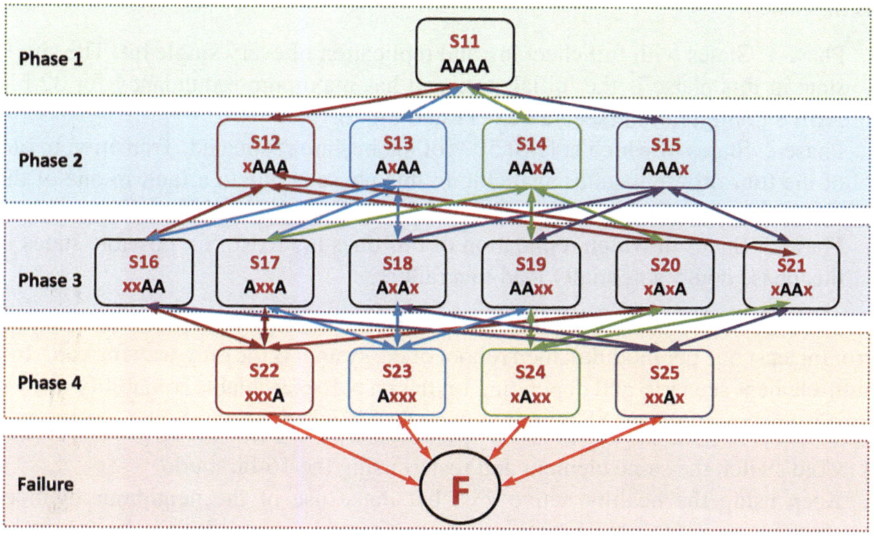

Fig. 6.16 32-bit degradation phases

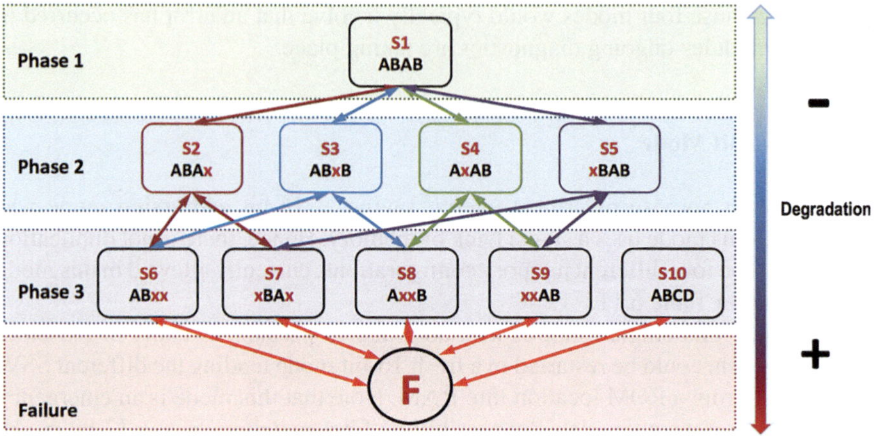

Fig. 6.17 16-bit degradation phases

The ten admissible states in 32-bit mode are given in Fig. 6.16. Further degradation is possible in 16-bit mode (Fig. 6.17). In terms of fault tolerance, the maximum redundant configuration is based on 32-bit addressing ABAB configuration. The transition between these states is completely dependent on soft/hard errors and the efficiency of recovery mechanisms. However, a voluntary transition between different states is also allowed.

In case of degradation, successful recovery mechanism could produce a transition from a degraded phase to an initial phase. The group of states or phases can be distinguished:

1. Phase 1: States with full checking and replication of every single bit. The single state in this phase is the initial state and has maximum redundancy for 32 bit: ABAB configuration (see Table 6.11 and Fig. 6.16).
2. Phase 2: States in which at least 50% of the bits are replicated. Transition to one of the four different states available in this phase is due to a fault in one of the memory modules of State 1.
3. Phase 3: States in which replication of bits does not exist. Six possible states in this phase could potentially lead to a failure.

If during the operation phase three working modules experience a third permanent error (at least one per module), then reboot of the system is the only way forward. In a multi-element scenario and depending on the resources available, reliability needed and current environment, SSW will have to decide whether to:

1. Cold switch the same element and restart using the 16-bit mode
2. Keep using the healthy active area but make use of the neighbour memory elements or,
3. Cold switch the execution to another element.

6.7.1 Graceful Degradation: Markov Analysis

Regarding computational capability and availability of resources, a system can be modelled as being in one of many possible states. The number of states would be large, if fine distinctions are made, or it may be relatively small if similar states are grouped together. Different events can force the system moves from one state to another, depending on resource availability and computational changes.

By quantifying the probability of state transitions, State-space modelling can determine the probability of the system being in each specific state; this can be used to obtain some parameters of resilience (reliability, safety, maintainability, safety, etc.).

The sum of all input and output transition probabilities of each state should be 1. The state of the system is characterized by the vector: $(S_0, S_1, S_2, S_3, \ldots, S_n)$.

A transition probability matrix has N states. On the tth time-step, the system is in exactly one of the available states q_t:

$$q_t \in \{S_1, S_2, S_3, \ldots, S_n\}$$

There are discrete time-steps, $t = 0$, $t = 1$, …. We are interested on how the system will behave after several time steps. Initial condition of the system: S_0. Given q_t, q_{t+1} are conditionally independent of $\{q_{t-1}, q_{t-2}, \ldots, q_1, q_0\}$ that is:

$$P\big(q_{t+1} = s_j | q_t = s_i\big) = P\big(q_{t+1} = s_j | q_t = s_i,\ \text{any earlier history}\big)$$

The current state q_t determines the probability distribution for the next state q_{t+1}. In order to model of the system, we do the following assumptions:

1. System starts in the perfect state
2. Only one fault can occur at a time.

In order to simplify, we make a first-order Markov assumption: we say that the probability of an observation at time n only depends on the observation at time n only depends on the observation at time $n - 1$. In a sequence $\{q_1, q_2, \ldots, q_n\}$

$$P(q_n | q_{t-1}, q_{t-2}, \ldots, q_1, q_0) \approx P(q_t | q_{t-1})$$

Using the previous assumption, the joint probability can be expressed as:

$$P(q_1, q_2, \ldots, q_n) = \prod_{i=1}^{n} P(q_i | q_{i-1})$$

Figure 6.18 presents Markov model of reliability for reconfigurable memory described above. The figure shows the transition probability in the case of four-module memory scheme with TMR plus a spare configuration. The circles in the

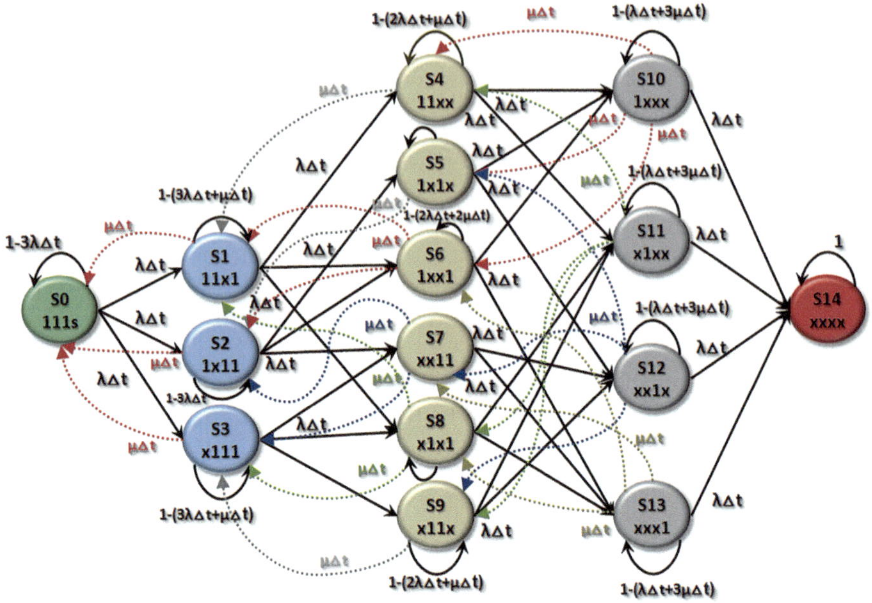

Fig. 6.18 Markov model for the ERA memory system

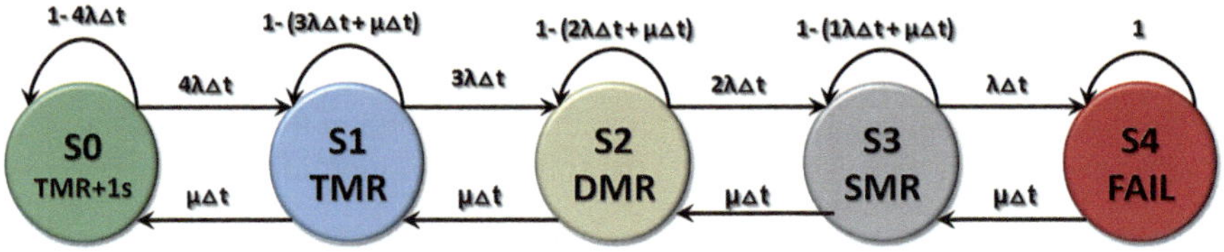

Fig. 6.19 Reduced Markov model for the ERA memory system

figure represent one of the 15 possible states. The arrow reflects transitions from source to destination state.

By merging states in the Markov model, a simpler equivalent model is created. Figure 6.19 represents such simplification, illustrating the probabilities of transition between the original TMR plus spare, a TMR, a DMR, a SMR and Fail states.

6.8 Implementation Constraints

As explained before in 6.2.1 taking into account the 16-bit instruction size and 32-bit word size of memory organization, at the time of compilation the compiler schedules memory loads on the first 16-bit instruction of a 32-bit two-instruction "packet". This way, memory loads and instruction fetches never occur at the same time. When the memory configuration is set as 16-bit words each read or write precede instruction execution. In the case of 32-bit memory configuration two instructions might be loaded from memory by one access.

6.8.1 Graceful Degradation: Solutions for Memory

Different solutions could have been implemented for ERA's memory addressing scheme. We consider physical addressing and relative addressing.

Following the first case, as shown in Fig. 6.20, the lowest 4 bits of 32-bit memory address would represent which modules are used to represent the lower and the higher 16 bits. The encoding would be as follows: Bit 2 and 3 correspond to M2 and represent the higher 16 bits.

However, by using this scheme, having the configuration of the modules physically mapped in program code would make reconfiguration difficult to implement. If binary code reflects which bank the program should run on, then in case of a fault that requires code to be transferred to another bank (bank switching), a translation would be needed.

This re-compiling of the program is highly unlikely possible in real time of program runs. This also affects recovery time and, in fact, it excludes the chances of recovery in real time.

Besides, if the contents of the memory banks are physically different, this will affect the hardware checker's complexity; the checker function would need to compare equal values in case of data comparison and different values in case of address comparison. Thus, the separation of concerns principle and a virtual memory approach are preferred.

By removing M1 and M2 (Fig. 6.18), the memory addresses used in a program code refer to a relative position, for instance, within a pair of modules AB. A reconfigurable memory controller (see Fig. 6.21) that links the relative address to the specific memory bank is used in a similar fashion to the translation of virtual addresses into physical addresses.

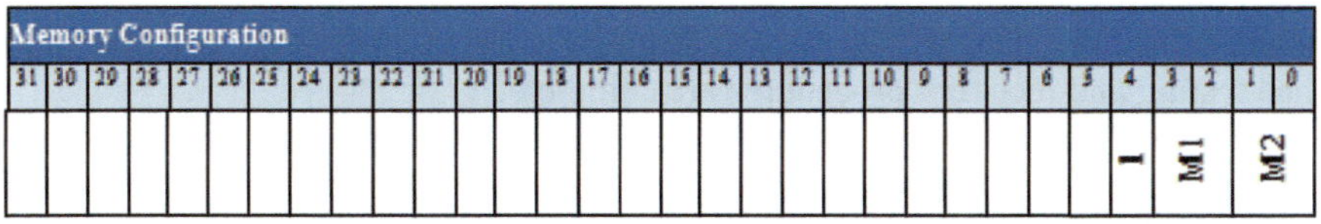

Bits	Name	Meaning
[0:1]	M2	The number of the memory module used for the higher 16 bit memory
[1:2]	M1	The number of the memory module used for the lower 16 bit memory

Fig. 6.20 Theoretical memory configuration with support of reconfigurability

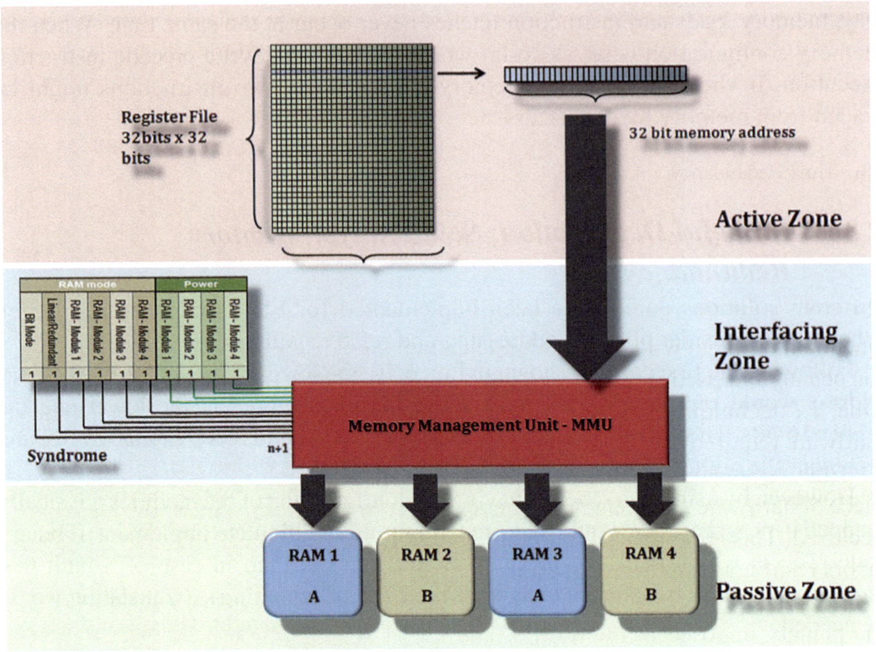

Fig. 6.21 MMU syndrome as a memory configurator

6.8.2 Interfacing Zone: The Syndrome as Memory Controller

As previously shown in Table 6.1 the T-logic ERA element performs configuration and reconfiguration of hardware by providing interconnection and dynamically excluding faulty components from the operational system.

The T-logic interconnector provides flexibility of application of memory elements (32- and 16-bit configurations) and at the same time helps in fault containment. This logic is used to form a hardware configuration scheme adjustable to the program requirements or when a hardware element itself (or architecture) detects hardware faults and thus can't be involved in further calculations.

Note that "isolation" might be temporary or permanent, subject to the element's "health". The final decision about permanent isolation of an element will take place when testing and recovery procedures are complete.

The four T-logic interconnectors, one for each memory bank, are physically included in the T-logic Management Unit or TLMU. TLMU (see MMU in Fig. 6.16) manages the connectivity of the memory, configures and reconfigures the working mode to a 16-bit single memory, 32-bit double memory with master/slave configuration or any of the 14 memory addressing schemes available (Tables 6.5 and 6.6).

Using the T-logic scheme memory elements could be isolated, switched off for power reduction or doubled in capacity when the maximum storage volume is required, addressing PRE-wise computing.

The Configuration and Power management flags of the syndrome describe the different states of the memory modules. Different values in the configuration area of the syndrome select the bank used and the mode.

The output memory lines of the processor determine a location within a memory bank, whereas the Configuration and Power areas of the syndrome specify which banks are to be used and in which mode.

One example of a possible memory configuration (State 1 Table 6.6) arranged by "T"-logic is presented in Fig. 6.16. The example reflects a 32-bit (Bit mode = 1) ABAB configuration with 2 × 2 modules duplicated (Redundant = 1) working in pairs.

By using this method we can increase the independence of software/hardware configurations for the PRE-purposes. Memory addresses within the code do not need to be arranged, as code integrity is a crucial requirement for safety critical systems.

Let's define the following scenario where it is required to switch data and code from modules 1 and 2 to module 3 and 4. Let's assume that the work mode is ABxx 32-bit, which is fast but not very reliable.

Assuming that the user (or an online fault detector manager using a fault model) requires a higher level of reliability, transfer from the current ABxx mode to an ABAB mode is required. An implementation of this example is based on the following algorithm (omitting the testing procedures):

```
!i!=!0;!
REPEAT!
!  i++;!
!  Starting!at!memory!location![0]!load!the!n!following!words!into!RF!(32x32)!
!  Change!the!value!of!the!syndrome!to!the!memory!mode!XXAB!
   Copy!the!n!previously!stored!words!from!the!RF!starting!at![0]!+!i*32*32!
!  i++!
!  Change!the!value!of!the!syndrome!to!the!memory!mode!ABXX!
UNTIL!EOM!(End!of!Memory)!
Change!the!value!of!the!syndrome!to!the!memory!mode!ABAB!
```

6.8.3 *Access to the Syndrome*

The design of the TLMU should allow the possibility of neighbour elements accessing the memory elements. Synchronization is then required to avoid several elements accessing the same memory at the same time.

Since the syndrome should also be accessible by other elements, it may be a good idea that the syndrome is located in this interfacing area. Having access to the syndrome via TLMU would save logic since only one synchronization element (TLMU) will be used.

6.9 Conclusions

1. A resilient architecture was proposed including the hardware as a structure and main elements, explaining how to handle efficient performance, reliability and energy use.
2. The principles of designed followed and the structural properties of active, passive and interfacing zones are introduced. Active zone of hardware was also described with emphasis on recoverability after malfunctions and implementation of checking schemes. A processor with a reduced instruction set and a careful introduction of redundancy including checking schemes and re-execution at the instruction level can provide higher and efficient reliability.
3. Reconfigurability of a real-time architecture at the system level was proposed and analyzed in the context of each zone. With regards to the interfacing zone, a new element (T-logic), as a basic unit of reconfiguration, and its different configurations were proposed. We analyzed and described how the flexibility of the T-elements has a positive effect in reliability and power-smart functioning of the system.
4. For system-level reconfigurability a new hardware element called *Syndrome* was suggested, to provide essential knowledge about hardware conditions. We showed the relation of this element with the active, passive and interfacing zones and how it can be used to implement generalized algorithm of fault tolerance. Functions of the *Syndrome* for reliability, performance and energy-smart functioning were described and explained.
5. Taking into account that memory use has, by design, a high impact on system reliability and power consumption, passive zone reconfigurability was analysed and described in detail, including the control of configuration and the phases of hardware degradation.
6. A Markov model of reliability for passive zone was developed and analysis indicates the reliability gain of the different schemes permitted by the proposed reconfigurable architecture.
7. System software support of testing and reconfiguration (dealing with system syndrome) was described. Shown that in combination of novel hardware architecture and system software all key properties of performance, reliability and energy-wise functioning can be improved.

References

Antola A, Erényi I, Scarabottolo N (1986) Transient fault management in systems based on the AMD 2900 microprocessors. Microprocess Microprogram 17:205–217
ARM Cortex A5. https://www.arm.com/products/silicon-ip-cpu/cortex-a/cortex-a5
ARM Limited (2005) ARM architecture reference manual. Technical report, ARM Limited
Gössel M, Ocheretny V, Sogomonyan E, Marienfeld D (2008) New Methods of Concurrent Checking. Springer.
Hennessy J, Patterson D (1996) Computer architecture, a quantitative approach, 2nd ed. Morgan Kaufmann
Hennessy J, Patterson D (2002) Computer architecture: a quantitative approach, 3rd edn. Morgan Kaufmann
Intel Corporation (2004) Intel pxa255 processor developer's manual. Technical Report. Order number 278693-002. Intel Corporation
Intel Corporation (2009) Intel 64 and 32 architectures software developer's manual, vol 1: Basic architecture. Technical Report 253665-032US
LEON3-FT SPARC V8—RTAX—Data sheet and user's manual. Aeroflex Gaisler AB, Goeteborg, Sweden, 1.1.0.8 edition, June 2009
Malyugin VD (1982) Polynomial implementation of a procession of Boolean functions. Dokl Akad Nauk SSSR 265(6):1338–1341
Schagaev et al (1989) Instruction set and their impact on modern computer architecture. Zarubeznya Electronica 7:8 (in Russian)
Schagaev I, Buhanova G (2001) Comparative study of fault tolerant RAM structures. In: IEEE DSN01. Presented at the IEEE DSN01, Goteborg
Schagaev I, Gutknecht J (2004) Software design for resilient computer systems. Springer. https://doi.org/10.1007/978-3-031-55139-0
Schagaev I, Kaegi T, Gutknecht J (2010) ERA: evolving reconfigurable architecture. In: 2010 11th ACIS international conference on software engineering, artificial intelligence, networking and parallel/distributed computing, London, UK, pp 215–220. https://doi.org/10.1109/SNPD.2010.40
Schagaev I (2008) Reliability of malfunction tolerance. In: Proceedings of the international multi-conference on computer science and information technology, pp 733–737. ISBN 978-83-60810-14-9. ISSN 1896-7094
SPARC International (1992) The SPARC architecture manual-version 8. Prentice Hall
Stepanyants A, Schagaev I (2001) Fault tolerant processor and its reliability analysis. In: IEEE DSN01, Goteborg
Verlag H (2009) ARM: Nachwuchs fr die die cortex-a-familie. World Wide Web electronic publication

Chapter 7
System Software Support for Resilience

Abstract Resilience of a system is not property any longer, it is a continuous process to prove that system is capable and operates as required... Our generalized algorithm of fault tolerance (GAFT) became a framework for pursuing resilience. GAFT might be efficiently implemented by hardware and system software holistically, pursuing all steps in concert—as one entity, together. This chapter presents essential features of software to support resilience: procedure of testing required to deal with different duration and character of fault, a monitoring of hardware configuration even in presence of hardware faults. We also present a scheduling of testing to provide required level of system integrity during its functioning. In the structure of operating system, we introduce a scheme and algorithm of control of hardware reconfiguration—so-called hardware monitor.

Keywords Resilience · Generalized algorithm of fault tolerance · Checking · Testing · Real-time scheduling of testing · Monitoring of hardware configuration · Operating systems

7.1 Introduction

In order for a system to be resilient and fault-tolerant several internal tasks should be active at every moment of system operation:

1. *fault detection and*
2. *fault determination*
3. *preparation of recovery.*

When fault is detected the *fault determination* starts and based on type of fault the reconfiguration of a system—if necessary—takes place. *Recovery*, in turn, is required when fault of the system detected, defined and software has to be returned to the state before fault manifestation.

These tasks can be easily decoupled, therefore, can be treated independently. The occurrence of a fault, in general, is a rare event. Thus, while error detection is on-going process, the error defining is essentially following, initiated action. Additionally, a

© The Author(s), under exclusive license to Springer Nature Switzerland AG 2026

I. Schagaev, *Resilience of Computer Systems*,
https://doi.org/10.1007/978-3-032-00938-8_7

recovery process is executed either by mentioned above fault appearance or activated as a result of regular maintenance procedures.

7.2 System Software Support of Hardware Checking

At first, testing of system condition. To prove that system functioning correctly some sequence of actions should be completed. Testing programs should be executed periodically to prove that system is working correctly.

Also, if there is a need to be sure that program was completed on hardware fault-free it is worth to run a test after useful program. In a bit more details: consider a sequence of tests and programs T and P within a system as in Fig. 7.1. The initial test T is executed before a given task execution guaranteeing HW consistency, i.e. it guarantees that there is no fault at time t_0 in the system.

However, if a permanent fault (e.g. a stuck bit) occurs immediately after the first test or during the program execution, it might be invisible for an arbitrary long time (latent period).

For periodic tasks, which are often used in control systems, we slightly adapt this scheme as shown in Fig. 7.2.

Nevertheless, what if a transient fault occurs during the execution of P? As mentioned before in Chap. 3, at least three cases may take place:

1. The effect of the fault lasts until P finishes, T detects the error, and the recovery mechanisms can be initiated and are able to restore normal functioning on time (detected recoverable error or DRE)
2. The effect of the fault lasts until P finishes, T detects the error, but the recovery on time is not possible (detected unrecoverable error or DUE)
3. The effect of the fault might not last until P finishes, and, therefore, T would not be able to detect the fault, which, in turn, would allow the fault to remain undetected, perhaps causing data corruption (silent data corruption or SDC).

Fig. 7.1 Ensuring HW integrity through program test execution

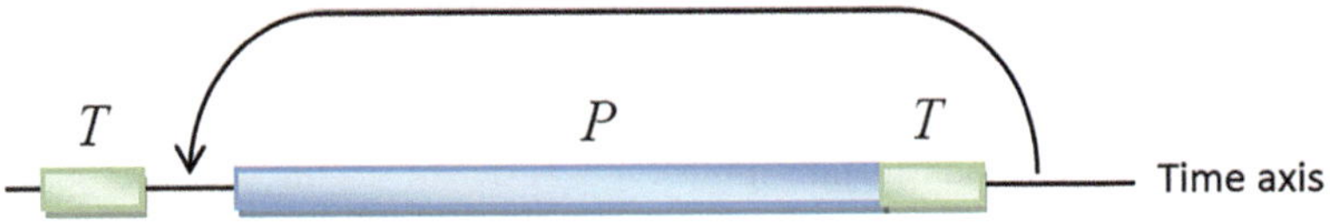

Fig. 7.2 Regular sequence of program execution with test of HW integrity to detect permanent faults

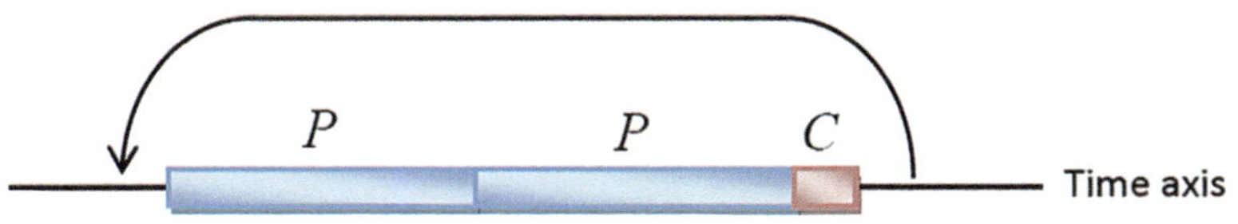

Fig. 7.3 Ensuring HW integrity through program test execution to detect transient faults

Thus, the corruption could go unnoticed (benign error) or could result in a visible error.

Transient faults can be detected by re-executing the same program P with comparison C of the result and the result state space.

Figure 7.3 illustrates this scenario. Note that for periodic tasks, the state of the program, which is used in the next computation as input data must also be compared, as transient faults might affect data that is no longer used in the current computation but in the next.

Permanent faults, however, cannot be detected with the comparison scheme alone, as they might affect both executions of P. That is, the scenario in Fig. 7.3 can only detect permanent faults whereas the scenario in Fig. 7.2 can only detect transient faults.

The combined power to detect transient and permanent faults is illustrated in Fig. 7.4 where C is used to detect transient faults and T to detect permanent faults. Assuming that C triggers an error but T does not, it is clear that a transient fault occurred. Another run of program P with comparison to the previous two runs, can identify the run where the transient fault occurred.

In the following analysis, we concentrate on the detection of permanent faults only and use only T in the analysis. The detection of transient faults can be considered as included in the following analysis if the double execution of P with following C as a whole is treated as task P in the following analysis.

A testing phase is required initially at boot up time to guarantee the correctness of the hardware and also a periodic test before and after the execution of a program. The applied tests might vary in depth (coverage), type of faults and the set of the tested hardware.

Every hardware component has typically at least one assigned test but might also have more than one that could differ on the implementation level. Software-based tests need a processor and memory for the test execution, even if a peripheral device is tested.

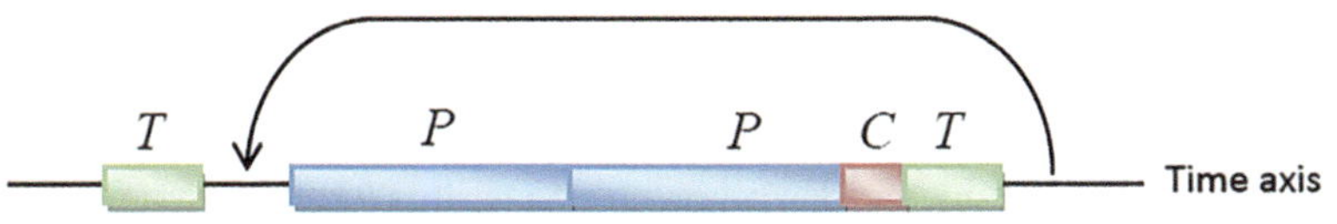

Fig. 7.4 Ensuring HW integrity through program test execution to detect transient and permanent faults

In order to guarantee that faults in other hardware components that are not subject of the test itself do not have an influence on the outcome of the test, the order of the tests must follow the *principle of growing core*:

> if a test of a hardware component u_i has implicit dependencies on another hardware component u_j, the test of u_j must be executed first.

If the resources needed by a task are known in advance, it is sufficient to run after the execution only the testing procedures of the accessed hardware resources (selective testing), again by using the principle of growing core. This way, the system stays fully operational even in the case of present faults in some hardware components that are not in use. Spare components can be used for the relocation of programs that were running on faulty hardware components.

Of course, it is also necessary—independent to this scenario—to arrange periodical test for the whole hardware as otherwise faulty spare components are considered as fully operational and might be used again in a subsequent reconfiguration process. Moreover, accumulation of faults can completely destroy a system.

A full hardware test also allows the system software to monitor the current full state of the hardware and take appropriate actions if necessary. If no spare components are available in the system, all programs depending on this component must be obviously terminated. If no essential program is affected by this component, the system can continue operating in a degraded mode.

For diagnostic and monitoring purposes, the results of the tests should be available for the software or even external systems. We propose therefore to organise the test results of hardware-based tests in so-called test syndromes before.

For every hardware component, for example the register file, AU, LU, internal bus or device controllers, the checking procedures present their result in the form of a syndrome to the software, indicating in binary form the state of the device. By grouping all syndromes together in one register, the software has a very effective way to check the integrity of the system. In case of a non-zero syndrome further analysis of the hardware conditions are required, especially when the malfunction duration is long.

Dependent on the used hardware checking scheme, it is not only possible to signal a fault to the runtime, but also provide more information to the runtime system to ease recovery.

If, for example, the testing schemes discover stuck bits in memory, it is sufficient to recover programs that access the affected location and not all programs that are using this memory module.

Device drivers could, for example, provide their own testing schemes for their respective device. Especially for devices, one could think of having a combination of hardware and software-based testing. I/O devices such as UARTs could effectively be tested by cross-connecting the input and output wires by very simple additional hardware logic and sending various bit patterns over this loopback connection.

Timely task completion in real-time systems is the key requirement, therefore, the testing overheads should be reduced as much as possible when and where possible.

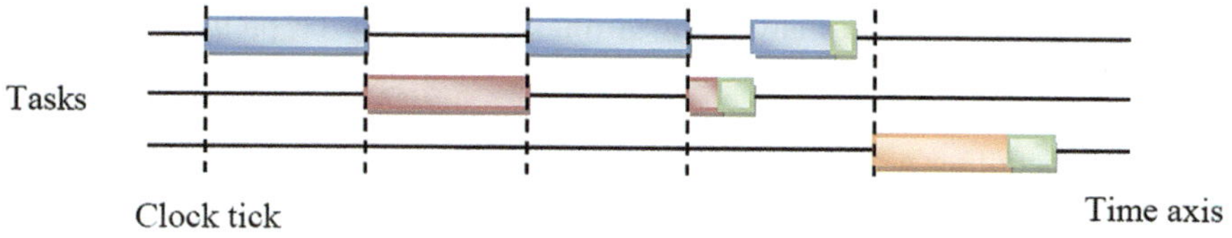

Fig. 7.5 Tasks and test combined

Figure 7.5 shows an example of three tasks with corresponding tests. Analysis of the checking process assumption in this case is based on a time slice scheduler, which distributes time slices to the running processes.

In this example, the processes run to completion and are called periodically by the scheduler. Three tasks are running, each with its own test (the green boxes) at the end of the task execution.

The test only checks the resources the respective process needs, which results in different test execution times. The task execution is only considered as successful if the test at the end of the task is successful.

If the test failed, the task is re-executed by using the same input data set as in the first try.

For diagnostic and monitoring purposes, the test results should be available to the system software with supportive manifestation in syndrome that unit or to external systems. Therefore, we propose to organize the test results of hardware in test syndromes.

Difficulties arise if the task performs I/O on hardware devices or communicates with other tasks.

There is no doubt that the hardware testing at the level of software, run-time system can be done at "spare time" of the system. In turn, while during "busy time", the procedures of testing and checking of hardware conditions must be different to avoid degradation of performance and task completion delays. Therefore, some schemes and impact on the system should be seriously analysed. Moreover, a combination of testing during idle time and busy time is also possible. In one of our previously published books (Schagaev and Gutknecht 2024) several chapters present complexities, analysis and solutions in full detail about it.

Still, in spite of prime goal of this book on hardware resilience it is required to read about adequate function of run-time system. Inside the runtime system a special support for new way of test scheduling is required. One way of it is presented in the next sub-chapter.

7.3 FT Scheduling

Scheduling in the event of faults is challenging, as the system must immediately diagnose the fault and define further actions that depend on the diagnosis outcome.

As faults can happen at any point in time, and cannot be predicted, dynamic scheduling is required to have the necessary flexibility in the system to react on the fault.

Static scheduling would not allow interrupting the currently running tasks and performing system diagnostic.

Before we introduce a scheduling algorithm that combines tasks and tests in the event of faults, we introduce some general guidelines we want to follow. Note that we assume a rate-monotonic scheduling algorithm.

This algorithm also includes the steps of GAFT. To illustrate the whole process, we first want to describe the process in words, as it is shown below:

1. *In case of a manifestation of fault, the process on the processor with the least critical load (idle, or the task with lowest priority) is pre-empted and the diagnostic routines are loaded.*
2. ***If** the fault affects hardware that is not in use, ignore the fault if the pre-empted process in Point 1 was a real-time task. If one processor does not run a real-time critical task, the system can use this processor to execute further actions (Point 4) If hardware is affected that is in use, continue at Point 4.*

   ***Otherwise** continue processing and delay the fault handling until later.*
3. *Perform GAFT steps B–E on one of the processors*
4. *GAFT steps F–J: Tasks that use the affected hardware and were running since the last hardware check are considered as faulty and recovery actions must be If the remaining execution time is not sufficient to execute recovered tasks before their respective deadline, consider executing alternate versions, depending on their execution times.*
5. *Reschedule tasks. Continue processing.*

Now, as seen above, the tests are scheduled by the scheduler and executed by one of the processors. If a test found a fault in the system, the test invokes the fault monitor, which then, in turn, informs the scheduler and performs the above-described actions. After doing this, the scheduler can reschedule the tasks and continue processing.

7.4 System Software Support for Hardware Reconfiguration

Run-time system interacts with hardware of the system through a hardware element named, across all our books, as a *syndrome*. All hardware checking schemes and regular control for every hardware component, e.g. register file, ALU, internal bus or device controller present their result in the form of a syndrome, accessible for run-time software. Syndrome indicates in binary form the state of the device.

Using a representative example of our ERA (evolving reconfigurable architecture) from previous chapter, we duplicate a figure of syndrome purposes, Fig. 7.6.

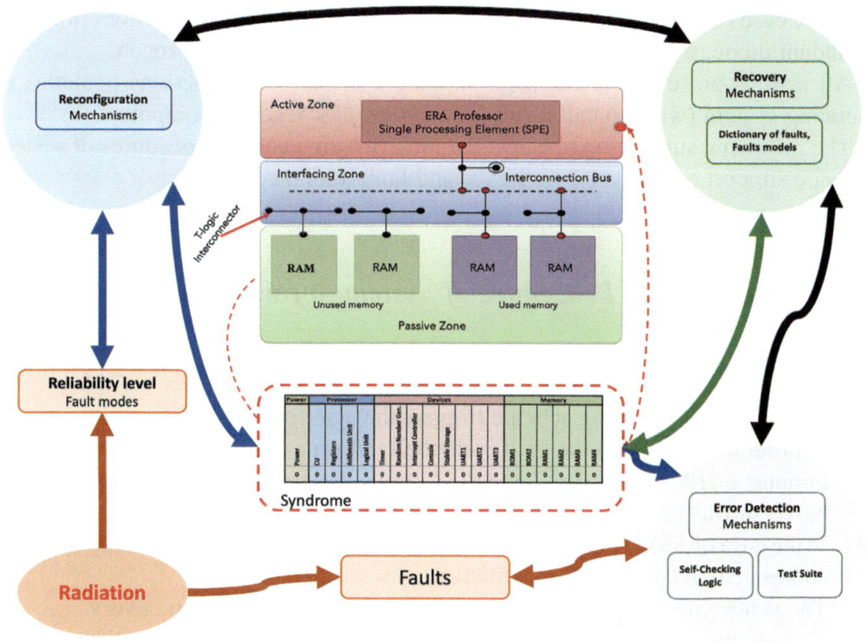

Fig. 7.6 Hardware syndrome and its purposes

Grouping of all signals of hardware conditions in the form of a syndrome—together in one register—enables the software to have effective way to check the integrity of the system.

Depending on the checking scheme used, it is not only possible to signal a fault to the runtime program, but also to provide extra information to ease recovery. For instance, let's define a scenario where the testing schemes discover stuck bits in memory due to a Hard Single Event Latch-up (SEL). Also, for this case—it would be sufficient to recover programs that access the affected location and not all programs that are using the affected memory module.

Device drivers could, for example, provide their own testing schemes for their respective device. Especially for devices, one could think of having a combination of hardware and software-based testing.

I/O devices such as UARTs could effectively be tested by cross connecting the input and output wires using very simple additional hardware logic and sending various bit patterns over this loopback connection.

In case of a detected hardware fault, the syndrome raises a hardware interrupt and SSW (system software) takes control of the reconfiguration process.

The whole procedure is almost identical in the case of software-based schemes detecting the fault. The only difference here is an origin—the interrupt that is raised is not hardware but software.

In the case of memory errors, if the current memory configuration does not use a redundant mode, software based checking is the only possible approach.

All actions presented at the page above are in fact new functions or duties of operating system (we also call it run-time system) for resilient computers.

The following subsection presents a summary form general procedures of system software support of hardware resilience and fault tolerance.

7.4.1 The General Procedure of System Software Support

The general procedure of software support during reconfiguration is listed as follows:

1. The hardware checking scheme triggers the syndrome interrupt.
2. In order to distinguish the fault type, the syndrome interrupt handler then either initiates a *HW BuiltIn SelfTest* (*BIST*) procedure of the device or runs a SSW based self-test.
3. In the case of SSW tests, writing different memory patterns to the faulty memory address can be used to derive the fault type.
4. The syndrome bit indicating the fault has to be cleared after recovery.
5. If after the test and recovery, the syndrome still shows the fault, the memory module is considered faulty.
6. The affected memory address is still present in one of the processor registers and based on the IRQ return address, the correct register number can be derived by decoding the memory instruction that triggered the fault.
7. In general, all software-based testing procedures that test memory must not use the stack (no procedure calls, no data pushed on the stack) until the proper functioning of the used stack locations is ensured.
8. In case of a transient fault, the event is logged and the program execution resumed. Logging the events is important as an accumulation of malfunctions in a module or a specified memory location could hint a potential permanent failure in the near future.
9. In case of a permanent fault, the current memory configuration is extracted from the syndrome, and the next degradation state is calculated according to the application needs and predefined degradation tables.
10. The new calculated memory configuration is written to the syndrome registers and the power of the faulty unit is removed. In some configuration transitions, the SSW has to adapt to the new situation and recover after excluding the faulty unit but before including the new one.
11. SSW clears the fault in the syndrome and resumes processing by returning from the interrupt.

Some of the presented transitions of hardware described in previous chapters—for our architecture in Chap. 6 need software intervention to fully recover from a permanent fault and to establish a new working software state. We show here a few situations where SW support is needed:

1. *Adding/Replacing a module of an already populated bank*: if a memory module of a redundant memory mode (DMR or TMR) that is suspected of presenting faults is replaced by another module, memory content must be replicated to the new module before it is included in the configuration. A small routine following the algorithm described previous chapters is sufficient to perform the copy without modifying the memory during the operation. Before performing that routine, SSW configures the syndrome to include a spare memory bank. After the copy operation, the spare module can be included in the working set.
2. *Failed module in the runtime system area*: The area of memory that contains by convention all runtime system data structures is critical for system operation. When the recovery procedures are unable to overcome the situation, if the module has a replicated pair in redundant mode the affected module is replaced by its replicated version via syndrome. However, if the module is not replicated, resetting the system either via a hardware watchdog or a software initiated power cycle is the option as a last resort. The BIST mechanism automatically identifies the failed module and configures another still working module to bank 1. The runtime system can then restart all critical applications.
3. *Fault in a memory module that is not replicated*: this case is the most difficult to handle as the software must adapt to the smaller available memory space.

Instead of graceful degradation, software can also decide to "upgrade" the system in terms of redundancy, i.e. going from a mode with less redundancy to a mode with higher redundancy.

The inclusion of a spare module corresponds to the first point in the list above; if an already used module is moved to another bank, software has to release all data structures residing on that module in case it is in non redundant use and repopulate it with data according to Point 1.

Intentional change of the operating mode to a less redundant mode is of course also possible, and needs no special software measures. As soon as the module is reconfigured to a free bank, the runtime system can start using it.

7.5 System Software Monitor of Hardware Condition

A hardware monitor, which is part of the runtime system is responsible for keeping track of the hardware state. For every hardware component, which is presented by the syndrome, the hardware monitor tracks the state in more detail than the system syndrome alone can provide.

It is also responsible for the execution of all software checking schemes and performs the actual hardware reconfiguration.

Thus, the hardware monitor by design, being a part of the runtime system, must have an access to the interrupt handler. However, this monitor should not be directly accessible by applications; only through drivers. Drivers of hardware are also a part of runtime system and can register state of hardware and initiate, if necessary, checking procedures for their respective hardware component.

When the system is turned on, the BIST procedures embedded in the system are executed. It runs tests on all devices, using the principle of growing core, to ensure the integrity of all devices. If a failure is detected, the syndrome sets the appropriate fault bits.

The BIST is also responsible to initiate the system to a predefined working state, i.e. the most reliable mode with all working available resources.

When the BIST finishes and passes control to the runtime system, the runtime passes control to the hardware monitor, which first mirrors the current state in software and then reconfigures the system according to the need of the program. As the syndrome might trigger an interrupt right after boot up, the syndrome interrupt handler has to ensure that the stack pointer is valid and if not initialise it.

From the "point of view" of system software every hardware component, which is managed by the syndrome, can has, as Fig. 7.7 shows, exactly one state.

This state diagram also shows all possible transitions between states, allowing the hardware monitor to reconfigure the system in a consistent way.

In fact, all of the above presented cases in the degradation scenarios where software intervention is required are clearly identifiable in Fig. 7.7.

Intervention is only required if the state transition goes from Stand-by to one of the active cases (marked in blue).

After boot up, all devices are either in state OFF or in one of the blue operation modes. As the BIST automatically configures the most reliable possible memory

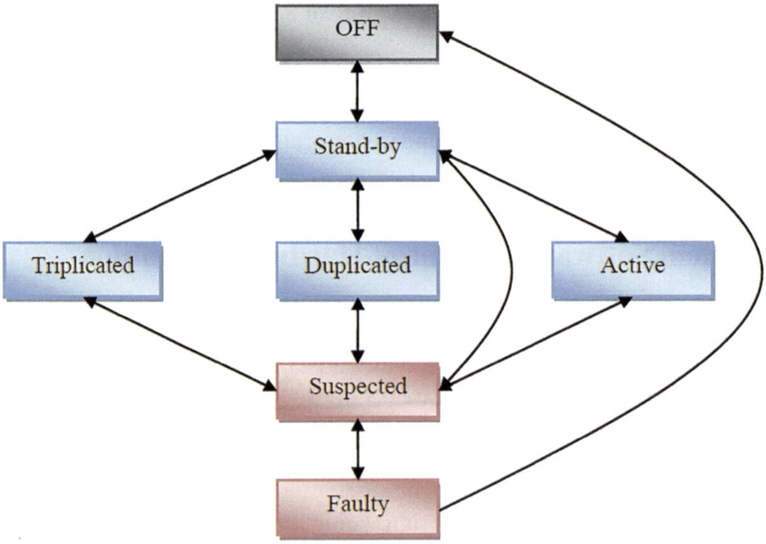

Fig. 7.7 Hardware state diagram

configuration, the initial states of all devices must be acquired by reading the syndrome. Here a short list of all possible states and a short description:

1. ***OFF***: the device is currently not in use, powered o_ and isolated for fault containment
2. ***Stand-by***: the device is powered on but not yet in use, i.e. in case of memory not yet assigned to a memory bank. In case of reconfiguration, all transitions go through this state.
3. ***Active***: the device is in use in a non-redundant mode. In case of memory, the memory module is assigned to a bank in linear non redundant mode.
4. ***Duplicated***: the device acts in duplicated mode.
5. ***Triplicated:*** the device acts in triplicated mode.
6. ***Suspected***: As soon as a fault in the hardware is detected, the state of the affected hardware component is set to suspected and the testing procedures are initiated to diagnose the fault. If a device is often in this state, this could be a hint that the device might fail in the near future. For reliability purposes, it might therefore be sensible to replace the component with a spare one.
7. ***Faulty***: dependent on the analysis outcome, the state is then set either to *Faulty* if a permanent fault was diagnosed or back to the previous state if it was only a transient fault. A device in the state Faulty is powered off.

The state transition diagram of Fig. 7.5 is not directly applicable to all devices. A memory bank, for example, should during the transition not go through "stand by" to make sure that the stored data in the attached memory modules are not lost.

As explained in previous sections of this chapter, periodical hardware checks should be performed on every single hardware component despite its state. This ensures that no hidden faults can stay undetected in the system and possibly spread.

Thus, to sum-up, from any workable "blue" state a transition exists to a state "Suspected".

Also, at "Stand-by" state—as currently not used devices—must be tested regularly, to avoid hidden fault accumulation.

It is even possible to revive faulty components, as for example environmental changes could allow a component to function correctly again.

7.6 Conclusion

System software support for hardware resilience was presented.

Various methods of testing and testing procedures were discussed in terms of implementation of hardware resilience.

Hardware fault detection, fault identification and recovery using system software procedures.

New system software scheme of scheduling of testing is described, providing real-time-ness of functioning.

A special hardware element—so-called syndrome was analysed regarding system software use for reconfiguration and recovery of system.

Graph and algorithm of hardware monitoring as a part of run-time system is presented, to achieve flexibility of system performance-, reliability- and energy-efficiency.

Reference

Schagaev I, Gutknecht J (2024) Software design for resilient computer systems, 3rd edn. ISBN 978-3-031-55138-3. ISBN 978-3-031-55139-0 (eBook). https://doi.org/10.1007/978-3-031-55139-0

Chapter 8
On Interruption

Abstract This chapter presents an analysis of interruption schemes for resilient computers. Starting first from concept of interaction, the specific features of interruptions for resilient system are discussed. Brief analysis of known architectures with interruptions is given. The way how to achieve resilience and efficiency of new interruption system using our T-logic scheme is proposed in details.

Keywords Concurrency · Interaction · Interruption · Mutual exclusion · Fault tolerance

8.1 Interruption as an Interaction

We read an Interruption—as interaction of entities without considering the inner state of each other.

Any interaction can be implemented using and applying the following categories and followed methods: informational, structural, time, further referred as I, S, T. Indeed, when an information is reaching a destination, it can change a state of entity. Both sides, bot entities can change their states initiated by interaction. This change might be considered as a flag: "I am ready" (to interact). And, again, structural, informational or time resource can be considered as an instrument of implementation of interaction.

Interaction use to be read as ordered, organized activity amongst entities. Initiation of interaction is assumed to be symmetrical—in other words—both sides have right to activate and use interaction for communication.

In turn, an interruption—on the contrary—is asymmetrical. It is an action of one entity to disrupt the other, or, if this process within the same entity, one part disrupts another at its own will.

Rigorously speaking, interruption is an interaction of entities without preliminary agreement to interact.

We can read as an interactive entity any external source of a signal or any internal part of the same entity, when a manifestation of a change of this part takes place.

© The Author(s), under exclusive license to Springer Nature Switzerland AG 2026
I. Schagaev, *Resilience of Computer Systems*,
https://doi.org/10.1007/978-3-032-00938-8_8

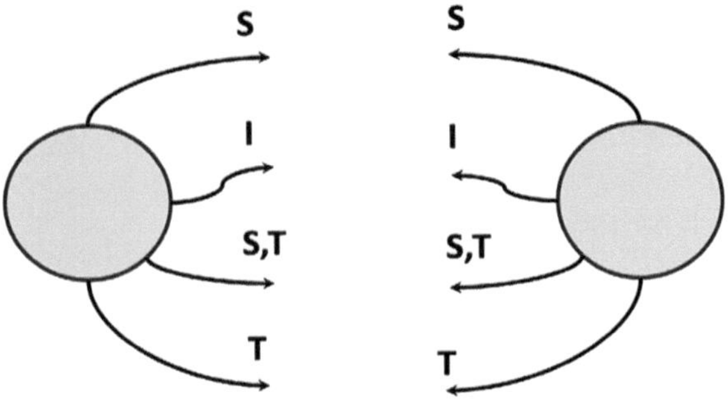

Fig. 8.1 Structural, informational and time options of interaction

Naturally, any alive system in the world should be able to handle interruptions, so are ours, human made systems. It is worth to see, again, through Fig. 8.1 that interruption can be implemented using structure, information, or time as a resource.

Note that combination of *structure* and *time* or combination of *structure* and *information* are allowed. Examples of this combination are well-known:

1. 1-bit register transmits any arbitrary length of data using, for example Morse code scheme structure—1 bit, information—unlimited, time linear dependent by information volume;
2. 8, 9, 12, 16, 18, 24, 32, 36, 64, 128-bit registers within our devices are capable with one tick of time (slotted time of work of a system);
3. 128, 256 bits registers are acting at every graphic processor.

Generally speaking, we are having a simple rule:

a total volume of information that can be delivered is defined by available hardware structure (registers, bus width, whatever) and time slot—a segment of time defined by technology of the available system.

Sometimes time slots are read as frequency of hardware clock in Hertz.

In principle, as it was mentioned already, the interactive entities might belong to different bodies. Also, two elements of interaction might be within the same body or entity (Fig. 8.2).

When a change inside of one entity needs to be served, one can apply the same scheme using various options—categories: structural, informational and time.

Defining interaction and further interruption, in fact, we have to extend Kant (1998) classification of interaction—*one-to-one*, *one to-many* up to *one-to-some*, *many-to-one*, and *some-to-one*. Regarding interruption we refer to three: *one-to-one*, *many-to-one*, and *some-to-one*.

Yes, methodologically our Graph Logic Model (Schagaev and Gutknecht 2024) provides excellent handling of entry and leaving points of and from entity. But handling of several active links still require some work.

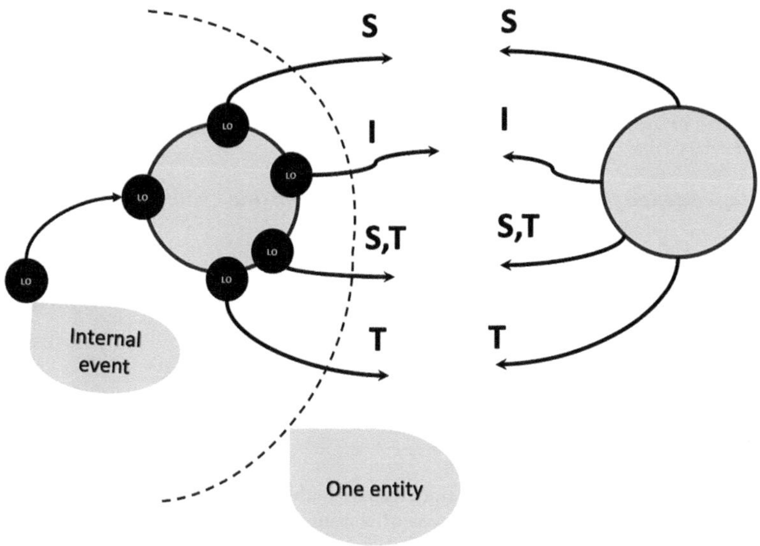

Fig. 8.2 Illustration of internal event causing interruption

8.2 Organisation of Mutual Exclusion

Here some principles of concurrency handling are presented as they were developed by computing society to achieve some decency in solving a synchronization problem.

For a successful communication (in philosophy—an interaction) of processes, the system must provide race-free condition; in other words, no matter which process comes first this time rule of entering interaction session must not depend upon ordering of execution.

Processes interests are different in their own time and interacting time, thus there is a reason to separate behaviour of the processes as a thing for itself and thing for us (more detail see E. Kant, Critic of pure reason) (Kant 1998). A section of the interaction of a process with others using resources of the system was called a critical section.

Critical section, therefore, has to have some obligatory and some recommended properties. Figure 8.2 presents only obligatory conditions. Above them, one might introduce desirable properties of synchronization system, as we have mentioned PRE-smartness, for example.

Thus, good concurrency system solution should be as good as Fig. 8.3 shows.

Let us consider several known solutions—well, fundamental ones, and check how these rules—obligatory and desirable are implemented.

This is from system level point of view. Let is briefly see what current hardware solutions are to implement interruption and how far are they from required conditions (Figs. 8.3 and 8.4).

- ***mutual exclusion*** *must be implemented - only one process must be permitted to be within critical section;*
- ***conflict resolution for simultaneous requests****, with one process to go through into the critical section; also called* ***liveliness*** *- if several processes request and entry one of them get access*
- ***independence*** *from processes demand from outside critical section*
- ***deadlock free*** *- no process to block each other*
- ***starvation free - bounding*** *- system does not let a process to wait forever; each process will get its time in critical section*

Fig. 8.3 Essential conditions to satisfy for any synchronisation

- ***efficient***; in other words do not consume substantial resources from the system or processes to arrange a process of waiting
- ***progress:*** *no process is forced to wait for an available resource -- otherwise very wasteful at a time;*
- ***do not implement busy wait*** - i.e. endless knocking the closed door is forbidden;

Fig. 8.4 Recommended conditions to satisfy for any synchronisation

Naturally, a priority of interruption is different—internal event service must prevail. This is because if, say, an internal fault in the system is not addressed immediately—as we introduced in the third edition of our Book (Schagaev and Gutknecht 2024): ALAP and ASAP—as local as possible and as soon as possible, we might face complete collapse of the system as a whole.

Also, an interruption as an example of concurrency handling at the level of hardware simply must not be using *time* as a palliative solution for synchronization within a system we build.

This alienation of the synchronization technology from synchronization concepts is crucial for the success of any concurrency handling and has to be analysed in more details.

Thus, the next steps here should be about known solutions and extracting the set of principles that must be pursued in the system design when concurrency appears to be solved.

Without synchronization of loosely tied processes, first, computers and other active entities were waiting for the completion of printer task, then perform memory reading (from a magnetic device called DRAM), and only then perform calculations. Thus, the slowest device was defining the overall performance of the whole system.

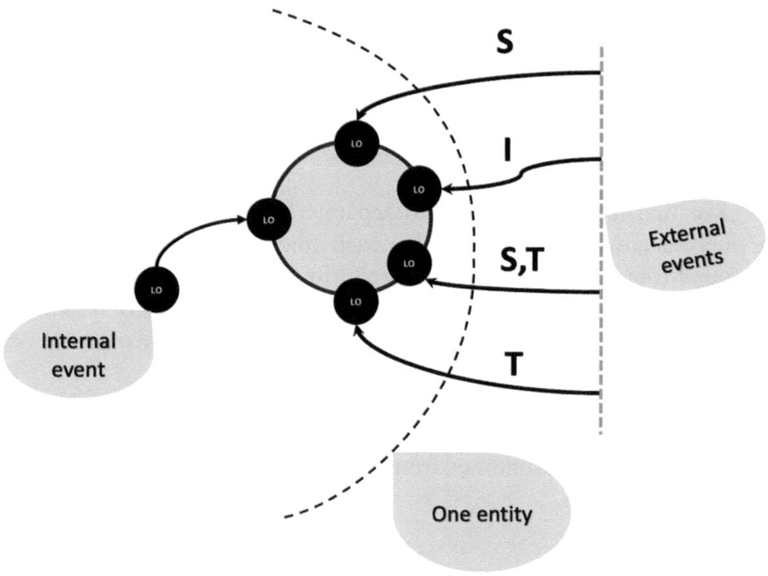

Fig. 8.5 Illustration of external requests—events causing interruption

Well, this is not right—as we described above—time is not a resource to play with making concurrency working.

Figure 8.5 also illustrates a concurrency problem—when various parallel processes came to the point of access to shareable resources.

Known hardware interruption mechanism is based either:

1. On round robin check of all entry points (potential interruption) giving an access to one at a time through structural support or mechanism, also known as vector interruption scheme
2. On segmented time slot for each entry when all time, say for n entry sockets divided amongst all defined sockets.

Well-known scheme of round-robin resource sharing as above was one of the first was implemented.

It is still in action on small local area networks and distributed computer systems (including multiprocessor systems).

In this solution *time* is used as a shareable resource, time allocated for synchronization-concurrency resolution is divided into slots, the number of equal slots is defined by the number of processes requested a slot of communication.

This scheme is "equal rights" scheme, where all processes are allowed the same amount of time and wait their turn next time around when all slots were given.

Interestingly, concurrency is a drawback of parallelization (explanations further follows) and both were designed to maximize the performance of the system.

Thus, the major drawback of round-robin synchronization is waiting time for each process—the whole round to be wasted.

Another drawback of round-robin synchronization is wastage of critical resource:

1. *No matter whether the process is in need or not the resource (slot of time) is allocated and, therefore ALL processes that requested synchronization will wait.*
2. Timing of executing the right to have a slot of communication becomes costly for all the rest.

It is clear that it is worth trying to separate concerns—personal interests and system interests are not always matching each other.

This round-robin approach was criticized, since Dijkstra time (early 60s of twentieth century) (Dijkstra 1965, 1968) as if time as a resource is in shortage one SHOULD NOT USE time as a resource to arrange coexistence of processes, their communication, interruption and completion.

It means that WHILE and similar waiting constructions should be used very carefully when it concerns a synchronization. Our personal opinion is that these types of construction should be prohibited from concurrency management and interruption implementation schemes.

8.3 Concurrency, Further Steps: Fault-Tolerant Interactors

We have been already describing our principles how to deal with a process (and hardware involved) when, due to a conflict of interest with others it attempts to access a so-called *critical section* of the system (Kant 1998).

The main innovation in this domain of human knowledge made by us, as far as we can see, is a concept of fault-tolerant semaphores (FTS). We consider FTS as an instrument to manage a process, while a process during its running has *reached a critical section* in the system.

This concept extends a condition proposed by Dijkstra during NATO school presentations in late 60s (Dijkstra 1968). In there all attention was about letting any process to have an access to the critical section.

The main difference of FTS is in widening Dijkstra and other renowned researchers' approach, is considering and managing a behaviour of a process *while it has reached a critical section.* And real problem of safety-critical systems is just about how to handle processes while they are already in critical section.

An approach of system software design for safety-critical system was sufficiently, we believe, explained in our book (Schagaev and Gutknecht 2024)—structurization of software with close adjustment of hardware, formation of recovery blocks with most the efficient searching of correct state of the system by applying our modified linear recovery algorithms. The efficiency of this algorithm is proven even mathematically.

And, if you carefully read Dijkstra assumptions, his synchronization assumes that non-critical section behavior of a process is not that much important, and system developer simply must turn to the design of procedures stepping in a crucial section. But! and this "but" is big:

The behaviour of a process in critical section CANNOT be ignored.

The assumption that eventually the process will leave a critical section is not strong enough: synchronization is about to maximize performance, to kick out a process out of CS by a timer 50–200 ms it looks like 10 million years later…

Thus, it is not a solution for real-time safety-critical applications and systems.

In this situation a question if system resilience and functioning of runtime systems remain unanswered:

What if a process that own critical section now "died" or "hang up" in there for an unlimited time?

All other processes will wait and die eventually. What can we do?

That is how an idea of fault-tolerant semaphore came to life.

Thus, the fault-tolerant semaphore means that any change of process condition due to internal fault must automatically lead to:

1. *release a critical section and*
2. *release all system resources that were used*
3. *signal the rest of the system about a faulty behavior of this process.*

Further, it implements fault-tolerant semaphore if we are using the algorithm of Fig. 8.6.

We will require a change of variable "turn" by reducing the number of processes that will be permitted to apply for access to the critical section after a faulty process emerges.

Thus, for N scheduled processes monitor of interruptions should reduce the number of "applicants" to $N - 1$. Also, we define turn as "other", that may take value either 1 or 2, while status out as 0, and status of competing as 1.

The status bits are initialized as "out". Another few hundreds of versions of the similar code are everywhere starting from Wikipedia and ending every second Ph.D. on the subject in the world and, therefore, there is no need to refer to any authorship here—this is public domain knowledge.

Fig. 8.6 Competing for a resource

```
status [i] := competing;
while (status [other] = competing)
{
if (turn = other)
        {
```

Note that in the case of mutual exclusion: Fig. 8.6, we have to assume un-interrupting execution of tests of global variable and setting of own variables.

This is supported by hardware instructions usually called "***test and set***" introduced in early 50s hardware architectures.

Mutual exclusion is implemented here because entering the critical section is taking place if and only if:

1. *either other process does not want to enter, or*
2. *other process wants to enter, but it is not its turn.*

A property called progress is achieved: both processes cannot wait forever in a While loop.

The process gets access to the critical section immediately if other processes do not want to enter, or other processes (matching turn) will eventually finish.

Starvation-free property is also fulfilled as one process after entering the waiting phase will wait at most one critical.

To avoid fast degradation of the system due to reconfigurability of hardware, in this case one has to (eventually, when the workload is not critical) to check the health of a process that was self-declared dead.

This is because, for example of natural radiation impact after which malfunctions in the real-time systems now lasts up to for 200–350 ms. Book (Castano and Schagaev 2015) has some chapters of that book might be useful to get details of reconfiguration handling.

This is known the duty of runtime system—to deals with synchronization. Implementation of the proposed approach and, specifically fault tolerant semaphores will improve the performance of synchronization of the system and, at the same time, makes system resilient—no matter which process fails.

An approach of FTS was presented to Dijkstra by a letter and his reaction was extremely positive, unfortunately he passed away the same year and collaboration on it was lost.

Since then, to our best knowledge, nobody neither proposed or implemented anything similar and this is a subject of further modernization of any runtime system for resilient computer systems.

Known Hardware Implementation of Interruption Support

As we have mentioned above, there are two known schemes—using time separators, and structure separators—in hardware it is called vector interruption scheme.

Daisy-chain, time slotting solution is about serially connected of all potential sources of interaction and interruption, connected serially on the bus within a system. It is also known as time-sliced scheme.

This solution assumes serial connection serially of units—all devices on a shared request line. This topologically set a priority and order—first in first served and the highest priority goes to the first in line. As such a first high priority interrupt is dealt first and then the second and so on as shown in Fig. 8.7.

An alternative and known solution is so-called vector organization of interruption in the system (Fig. 8.8).

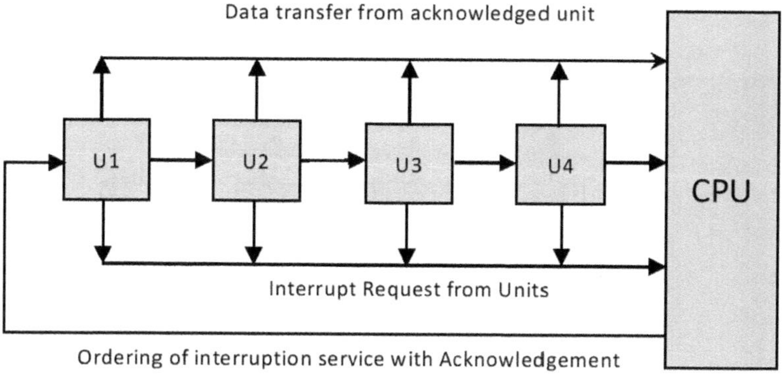

Fig. 8.7 Sequential ordering of interrupt units

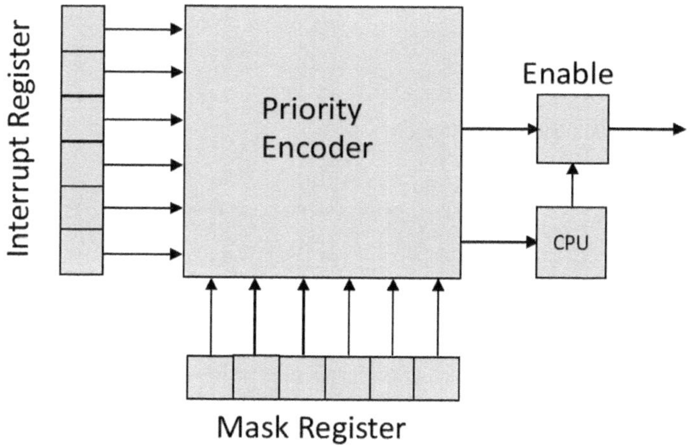

Fig. 8.8 Vector-ordering of interruption

Here, a priority interrupt in hardware captures the interrupt request from various external sources and upon the programmable mask sends the request with the highest priority to the CPU. Using control lines for interrupt enabling or interrupt acknowledge signal coming from the CPU a synchronization of the process within the system.

Sometimes this scheme is called vector interruption scheme as all sources of interruption can claim an acknowledgement independently and in parallel. In turn, changing the mask gives some flexibility for interrupt handling within a system.

Within our mentioned books the following hardware scheme for interruption was proposed, taking into account PRE-smart properties. PRE here as before stands for Performance, Reliability and Efficiency of proposed scheme. Figures 8.9 and 8.10 give some details. Our construct is called "T-logic", was created for the purpose of

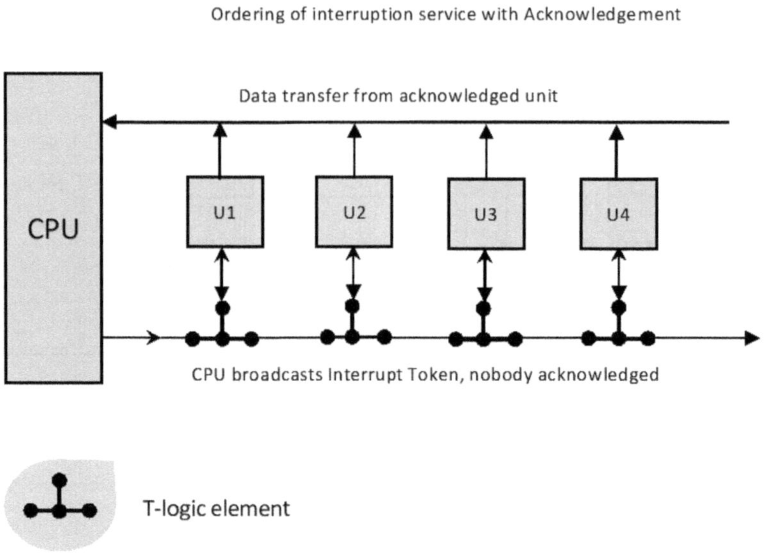

Fig. 8.9 T-logic instalment for interruption

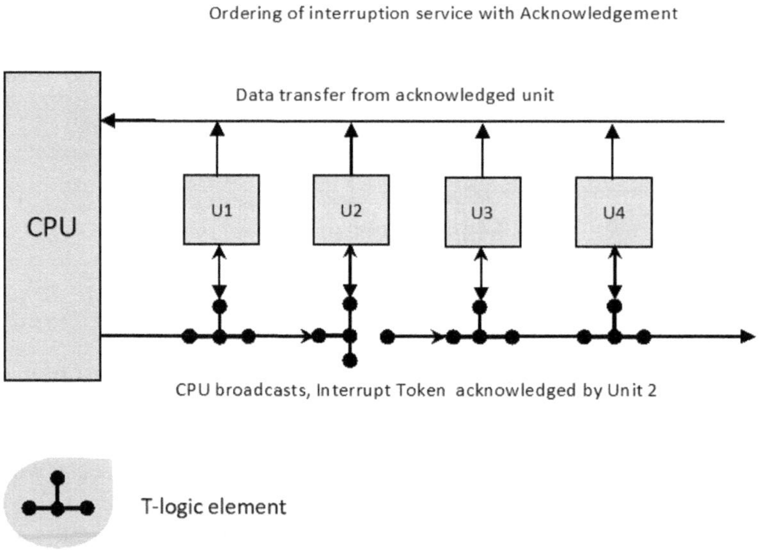

Fig. 8.10 Unit 2 takes a system signal

reconfigurability of hardware to address the need of tuning hardware for program tasks, isolate faulty component and make dormant not required components for saving energy when required.

Detailed description of T-Logic scheme follows in Appendix 1 for this chapter, while Appendix 2 presents an example of source code for possible scheme to implement T-logic functions.

For the principle, flow Fig. 8.9 illustrates that in the line of interruption permission signal or simple strobe T-logic element is connects line from the system and hardware Units—1–4. To make a request for a service unit should raise a flag—also known as a semaphore that separate these requests from other "desires" of other units. T-logic for this purpose needs to execute the turn as Fig. 8.10 shows.

Naturally, the rest of the units will be isolated from the system line while Unit 2 occupies it. Usual techniques with this kind of system bus organization initiate next cycle and slot of time with allowances to get a connection from the beginning, for all possible agents again.

T-logic installment, in turn, enables a system to continue marsh route a token with permission from the last active Unit—Unit 2 on our figures. Figure 8.11 illustrates this.

The other advantage of using T-logic scheme within a system based on possibility to exclude permanently or temporarily any unit from interactions. As it was mentioned before, this switching off might be done for the purposes of reconfiguration, isolation of a faulty component or Unit, or energy saving. Figure 8.12 demonstrates it with Unit 4.

This feature is crucial for implementation of fault-tolerant semaphore without which the communication of any parallel processes is not actually correct.

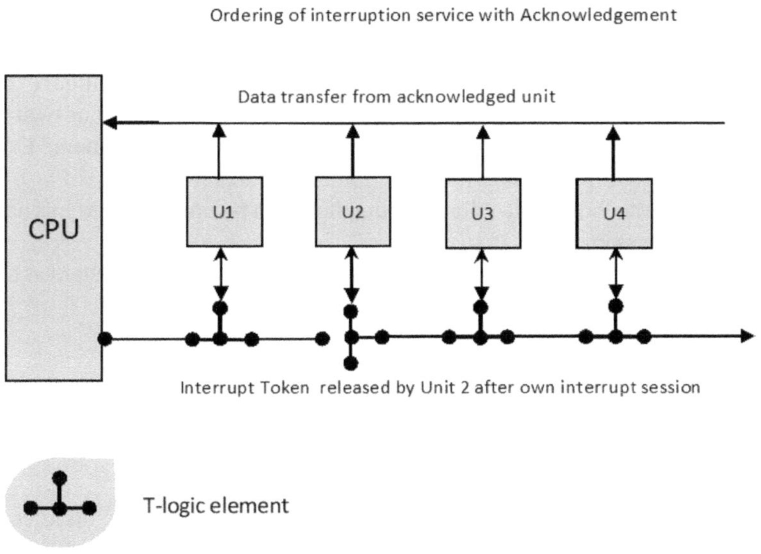

Fig. 8.11 Unit 2 finished using a bus, released it

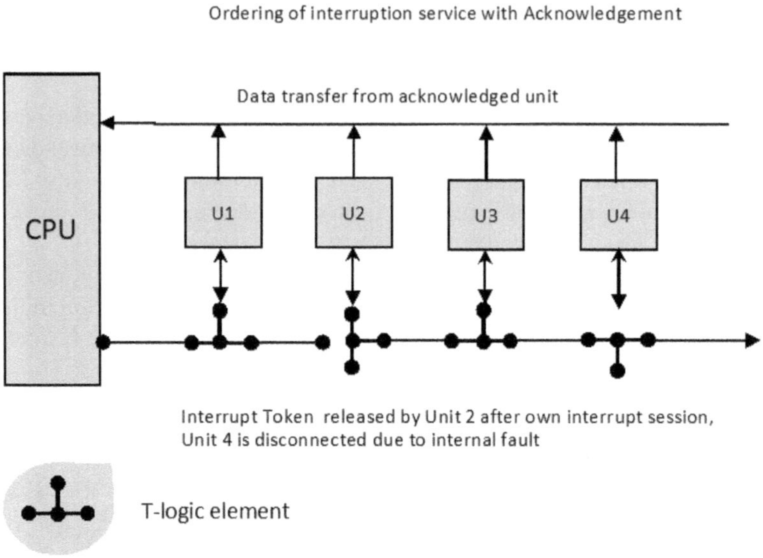

Fig. 8.12 Unit 4 is disconnected from the system by T-logic element

8.4 Conclusion

1. From the top level of systems actor interactions, we elaborate what kind of resources and solutions are required to make interruption efficient.
2. We suggest that timers and other options for using various delays is not a good practice to make interruption efficient.
3. Semantics of interactions in terms of available resources and synchronisation is discussed going down to necessary schemes and algorithms of mutual exclusion.
4. Starting from classic conditions for synchronisation and using our own models we design so called fault-tolerant semaphores that, when implemented, exclude process hang while within critical section of its execution.
5. We demonstrate how fault-tolerant semaphore can be implemented in a hardware of resilient system.
6. Shown that fault-tolerant semaphores increase resilience of the system as a whole.
7. Implementation variants are presented in Appendixes.

Appendix 1: T Logic True Table

T logic was suggested as a part of ERA (embedded reconfigurable architecture) to support fault tolerance in the ERA architecture. This is achieved by disconnecting faulty components while they are being repaired of diagnosed or if failed completely to isolate them. The current results show that only the vector A to B is working as a bidirectional bus and can also branch out to either clockwise or anti-clock-wise connections, i.e. C or D.

The T-logic is made up of 6 ‘IO buffers’ and 4 ‘NOT’ gates. The input ports ‘dirA, dirB, dirC and dirD’ as well as the switch ‘sw’ are the control lines of the IO buffers. For the ability of the T-logic to connect to other component, ports or buses the tri-states characeristics of the ports ‘biA, biB, biC and biD’ are required. This means that these ports can be made as either an input, output or a third state with high impedance. The control line can be minimised as seen from the Truth Table dirA is invertered to dirB with an added ‘Not’ gate to the confirguration.

Truth Table for T-Logic Fault-Tolerant Connection

(1)							Bus		Branch	
		A → B → C					Input	Output	Output	
		sw	dirA	dirB	dirC	dirD	biA	biB	biC	biD
		1	0	1	0	0	0	0	0	Z
		1	0	1	0	0	1	1	1	Z
(2)	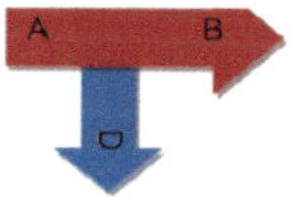						Bus		Branch	
		A → B → D					Input	Output		Output
		sw	dirA	dirB	dirC	dirD	biA	biB	biC	biD
		0	0	1	0	0	0	0	Z	0
		0	0	1	0	0	1	1	Z	1
(3)	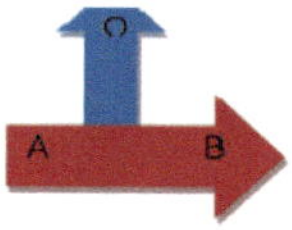						Bus		Branch	
		B → A → C					Output	Input	Output	
		sw	dirA	dirB	dirC	dirD	biA	biB	biC	biD
		1	1	0	0	0	0	0	0	Z
		1	1	0	0	0	1	1	1	Z
(4)	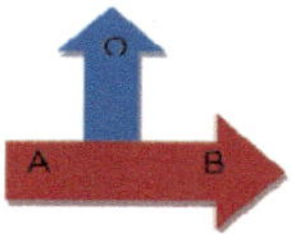						Bus		Branch	
		B → A → D					Output	Input		Output
		sw	dirA	dirB	dirC	dirD	biA	biB	biC	biD
		0	1	0	0	0	0	0	Z	0
		0	1	0	0	0	1	1	Z	1

Appendix 2: T Logic True Table Source Code

T-Logic

Date: August 29, 2012 Tlogic.v Project: test

```
/*
This module represent the implementation
of the T-Logic element
Last update 6/8/12
*/
module Tlogic(state,back,front,left,right);

input [1:0]back,left;
output [1:0]right,front;
input [2:0]state;
reg [1:0]front,right,center;
//input Fin;
//output Fo;

always@(*)
begin
front<=2'b00;
right<=2'b00;
center<=2'b00;
case(state)
3'b000:begin//BRF
front<=center;
center<=back;
right<=center;

end
3'b001:begin//RL
right<=center;
center<=left;
end
3'b010:begin//FL
center<=left;
front<=center;
end
3'b011:begin//RF
right<=center;
front<=center;
end
3'b100:begin//RFL
right<=center;
front<=center;
center<=left;
end
3'b101:begin//BR

center<=back;
right<=center;

end
default:begin
end
endcase
end
endmodule
```

References

Castano V, Schagaev I (2015) Hardware design of resilient computer system design, 2015th edn. Springer. ISBN-10: 3319150685, ISBN-13: 978-3319150680

Dijkstra EW (1965) Solution of a problem in concurrent programming control. Commun ACM 8(9):569. https://doi.org/10.1145/365559.365617

Dijkstra EW (1968) Cooperating sequential processes. In: Genuys F (ed) Programming languages: NATO advanced study institute. Academic Press, pp 43–112. Reprinted in Brinch Hansen P (ed) (2002) The origin of concurrent programming. Springer, pp 65–138. https://doi.org/10.1007/978-1-3457-3472-0_3

Kant E (1998) Critic of pure reason. Cambridge University Press. https://doi.org/10.1017/CBO9780511804649.003

Schagaev I, Gutknecht J (2024) Software design for resilient computer systems, 3rd edn. Springer. ISBN-13: 978-3031551383

Chapter 9
Implementation: Hardware Prototype, Simulation and Testing

Abstract We present our own prototype of a resilient computer designed and developed from scratch. Format of instructions, the instruction set itself is presented and explained. Special features of instruction flow are explained. Assembler and essential features of its implementation are explained and demonstration how it works using prototype algorithms.

Keywords Evolving Reconfigurable Architecture (ERA) · Instruction set · Instruction execution · Hardware prototype · Simulator of architecture · Assembler

9.1 Introduction

There is always a gap between good theory and concept, analysis as well as "what we design" and "what we can implement".

While analysis of the technological domain and trends as well as a new theory of resilient systems were presented in the previous chapters, as a next step, there is a need to demonstrate "proof of the concept" and do-ability of the proposed ideas on our existing hardware prototype.

Thus, the implementation efforts can be summarized as two-fold:

1. Development of hardware prototype
2. Debugging of existing manufactured ERA prototype;
3. Simulation of hardware for deeper analysis;
4. Design and development of software simulator for the developed hardware

As we see the further steps of development the hardware prototype must work properly, being able to execute the basic functions that will be further incremented, so that the system is prepared for industrial use.

In applications, ERA will work as stand-alone system using interface with PC or Mac to upload and run programs in a form of bit stream as we use FPGA Altera for this.

System software and programs for ERA also have to be developed and, therefore, there is a need for a software simulator for the ERA architecture.

© The Author(s), under exclusive license to Springer Nature Switzerland AG 2026

I. Schagaev, *Resilience of Computer Systems*,
https://doi.org/10.1007/978-3-032-00938-8_9

As long as ERA is designed for embedded and safety–critical applications, we need to have a simulator as close as possible to the architecture and ERA's binary code.

Therefore, all our efforts in this chapter are around the simulation of the instruction set and the hardware elements.

9.2 Instruction Execution

Figure 9.1 shows the execution flow of the proposed microprocessor. As we mentioned in previous chapters, primarily in 6.2.1, the execution steps are similar to other RISC processors and, since there is no pipelining, all steps of one single instruction are executed within one memory cycle.

The fetching step loads an instruction from main memory, storing it into the Instruction Register (IR). Since every instruction is 16-bit wide this step is required every second instruction. Decoding and Execution of the Instruction follows.

After execution of the first instruction, the second instruction from the Instruction Register (IR) can be executed without access to memory. This eases the speed gap between processor and memory and reduces their dependency.

Finally, storage of the result takes place if the instruction execution has affected the content of registers, processor flags or any other processor state. The processor has two internal fetching states (F1 and F2) that are required by the memory controller.

Again, simplicity is not only applied to the processor but also to the memory controller. Both are designed to avoid possible stalling due to pending memory operations. This can be achieved by interleaving instruction fetches and memory operations.

The fetching step always loads two 16-bit instructions from memory into the internal IR. In sequential instruction execution, the compiler can schedule memory instructions in every second instruction slot where no instruction has to be loaded by the processor.

An example of this notion is illustrated Fig. 9.2, which is an extended version of the instruction execution flow in Fig. 9.1.

In the F1 state, the processor fetches an instruction from memory; therefore, since the memory unit is busy during this cycle, it cannot execute at the same time an instruction that involves memory.

Only when the processor is in the fetching state F2, the processor is able to execute a memory instruction. It is the compiler's responsibility to schedule the instruction in the proper order.

We choose to simplify the hardware design at the expense of adding complexity to the compiler. By doing so, we reduce the amount of redundancy and therefore increase system reliability. If the instruction executed is a branch instruction the processor is switched automatically to the F1 state.

The reason for that is that the memory controller can only load 32-bit aligned addresses and therefore, the jump destination locations must also be 32-bit aligned.

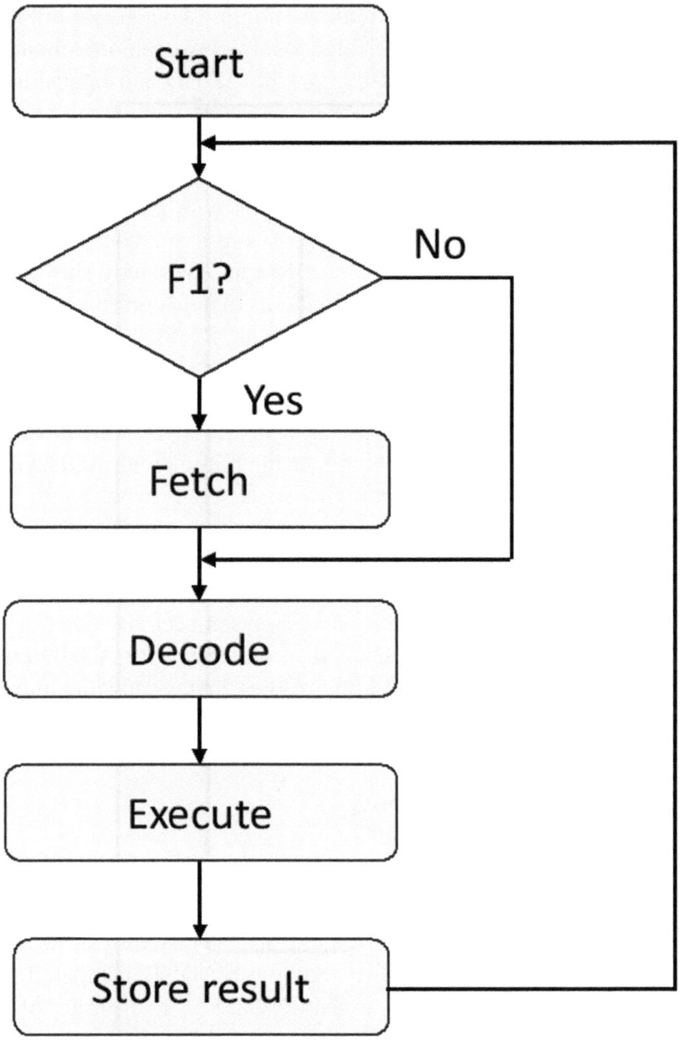

Fig. 9.1 Simple version of the prototype's instruction execution

All these arguments and presented factors force the compiler to fill the memory gaps with NOP instructions.

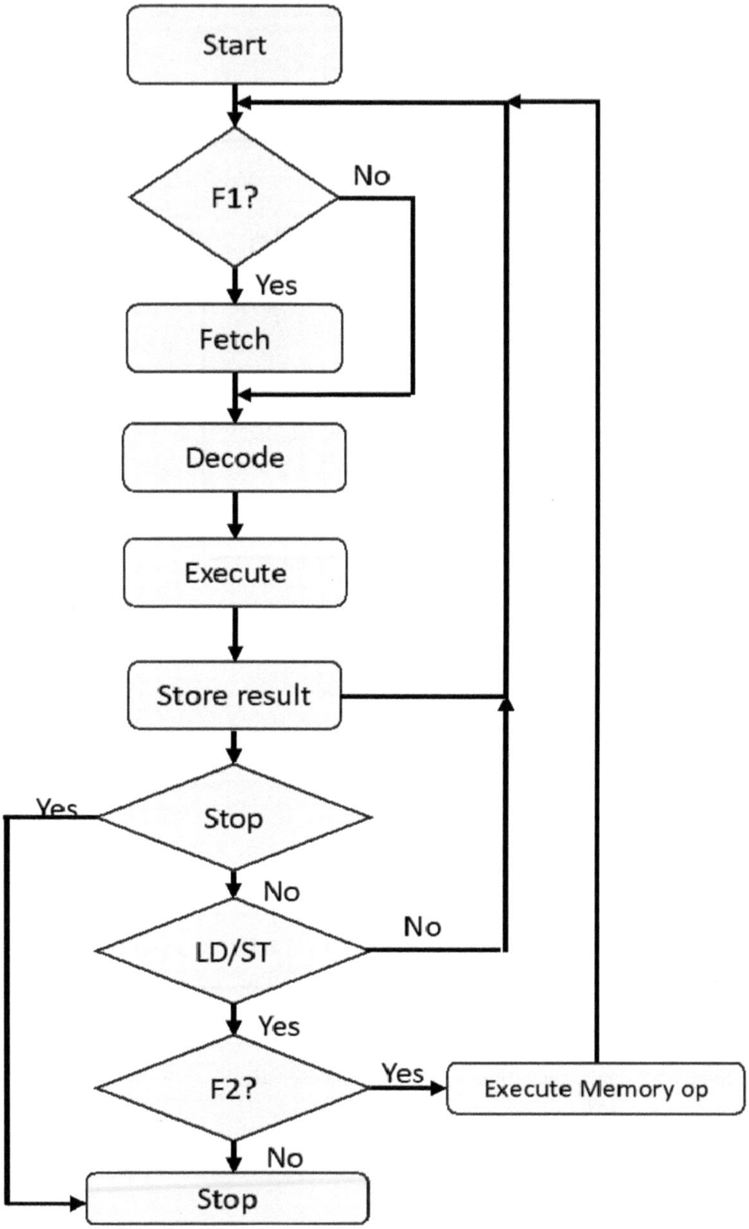

Fig. 9.2 Instruction execution (extended version)

9.3 Instruction Set

As mentioned before, in the ultra reduced instruction set employed, each one of the 16 instructions is encoded into 16 bits and only two of them are memory instructions (load/store).

The instructions are designed as two operands ones. They expect exactly two arbitrary registers as input, and save the result of the operation in the first register, thus overwriting one of the input values.

The compiler has to be designed with awareness that the content of the first register is overwritten. Again, we increase the simplicity of design at the expense of compiler's complexity. The impact of the size of the register file on overall performance of processor is a question of further research.

Figure 9.3 illustrates the instruction format divided in four different areas. Bits 15 and 14 (in red) indicate the format of the operation, which could be 8-, 16- or 32-bit. Bits 13 to 10 (in grey) contain one of the 16 different operation codes. Bits 9 to 5 (in blue) and 4 to 0 (in green) contain the first and second operand, which could be any of the 32 general purpose registers in the *RF*.

The following Table 9.1 lists the current ISA with a short description of every instruction together with their assembler representation. The current architecture comprises 7 control, 5 logic and 4 arithmetic instructions. The most entries in Table 9.1 are self-explanatory, when we feel necessary we add some elaborsation.

A special case of code operation is Opcode 0 which in combination with the format represents four different instructions: *STOP*, *NOP*, *TRACE* and *RETI*. *TRACE* is used for debug purposes and *RETI* is used to exit an interrupt handler.

The *LDA* is another special instruction that loads a constant to the specified register. The next aligned 32-bits aligned after the active program counter store the constant to be loaded. Placing two instructions into one 32-bit word increases code density and performance.

As explained earlier, the processor executes the first instruction on the left before executing the second one on the right. Since the program counter has the same value for both instructions it would be problematic to jump directly to a second instruction, which is not 32-bit aligned and has no unique address.

Therefore, it is responsibility of the compiler to insert *NOPs* at the right places to prevent cases where instruction reordering fails to fill the gap.

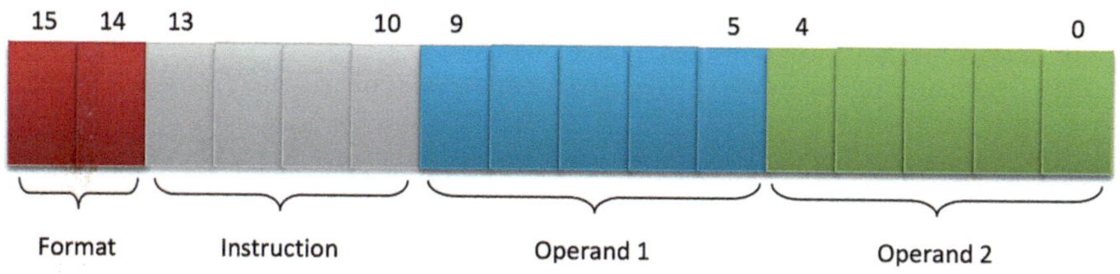

Fig. 9.3 Instruction format

Table 9.1 Instruction set explained

Instr. name	Format code	Instr code	Op1	Op2	Description
NOP	01	0000	Ignored	Ignored	Execute no action except increasing PC
STOP	00	0000	0	0	Stop instruction execution
TRACE	00	0000	$Ri > 0$	$Rj > 0$	Output *Ri* to debugger operand 1 or operand 2 must be > 0
RETI	11	0000			Return from interrupt
LD	11	0001	*Ri*	*Rj*	Load 32-bit memory at address *Ri* into register *Rj* (*Rj*: = **Ri*)
LDA	11	0010	Ignored	*Rj*	Load the value from the next 32-bit word ref to PC and store it in *Rj (Rj: = constant)* Operand 1 is ignored
ST	11	0011	*Ri*	*Rj*	Store content of register *Ri* to the memory at address *Rj (*Rj: = constant).* Operand 1 is ignored
MOV	XX	0100	*Ri*	*Rj*	Move content of *register Ri to register Rj (Rj: = Ri)*
ADD	XX	0101	*Ri*	*Rj*	Arithmetically add the content of *Rj* to the content of *Rj* and store the result in *Rj Rj:* = *Rj* + *Ri*
SUB	XX	0110	*Ri*	*Rj*	Arithmetically subtract the content of *Rj* to the content of *Rj* and store the result in *Rj Rj:* = *Rj* – *Ri*
ASR	XX	0111	*Ri*	*Rj*	Shift the content arithmetically one bit right and store in *Rj*
ASL	XX	1000	*Ri*	*Rj*	Shift the content arithmetically one bit left and store in *Rj*
OR	XX	1001	*Ri*	*Rj*	Perform a bit wise logical OR of register *Ri* with register *Rj* and store the result in *Rj*
AND	XX	1010	*Ri*	*Rj*	Perform a bit wise logical AND of register *Ri* with register *Rj* and store the result in *Rj*
XOR	XX	1011	*Ri*	*Rj*	Perform a bit wise logical XOR of register *Ri* with register *Rj* and store the result in *Rj*
LSL	XX	1100	*Ri*	*Rj*	Shift the content of register *Rj* logically one bit to the left and store the result in *Rj*
LSR	XX	1101	*Ri*	*Rj*	Shift the content of register *Rj* logically one bit to the right and store the result in *Rj*
CND	XX	1110	*Ri*	*Rj*	Arithmetic comparison of *Ri* with *Rj* and store the resut in *Rj*
CBR	XX	1111	*Ri*	*Rj*	Jump to address in *Rj* if *Ri* is nonzero and save PC in *Ri*

Table 9.2 Bit mask for conditions

Relation	Bit mask
$<$	010
$\leq$	110
$=$	100
$\geq$	101
$>$	001
$\neq$	011

The *CND* instruction performs an arithmetic comparison of the Registers *Ri* and *Rj* storing the result in *Rj*. The functioning is similar to other platforms: the comparison involves checking of three conditions, saving these as flags in the first three bits of *Rj*:

1. Bit 0: $R1 > R2$
2. Bit 1: $R1 < R2$
3. Bit 2: $R1 = R2$

Table 9.2 shows the relations of the comparison operations and their corresponding bitmasks. By applying an appropriate mask to the flags, the result for every possible comparison operation can be used as an argument in a conditional jump or saved as a Boolean value.

At the moment there is no support for unsigned operations. All arithmetic operations treat the values in the operands as signed values. All instructions accept the same register for both arguments with the exception of conditional jump instruction.

9.4 ERA Hardware Prototype

Figure 9.4 presents the current custom hardware prototype of the ERA device. The prototype is provided with two flash-based ROM modules 128 Mb each (8 Mb $\times$ 16 bit), with replicated bootstrapping firmware and operating system software.

The proposed memory scheme may be regarded as a collection of 4 blocks, 16-bit wide, with identical size of 1 Mb each (16 $\times$ 64 k). Using 16-bit memory modules instead of 32-bit memory modules increases reliability and reduces when necessary energy required for execution.

It is worth presenting our first Fault Tolerant RAM implemented initially in a DIP package tested and applied since 1989, Fig. 9.5. Explanation of reliability and malfunction tolerance of this device was made in previous chapters, by rewriting of a corrected output to all inputs of the memory in fact prolong working life of this memory at order of magnitude.

In the prototype device of Fig. 9.4, this malfunction tolerant triplicated RAM due to its extreme reliability was implemented within Altera Cyclone II.

As Fig. 9.5 illustrates, FTRAM is 32 K 8-bit, Industrial, triplicated.

Fig. 9.4 ERA prototype board

Again, reliability of a devices is increased by means of added working states and configurations.

Energy-wise operation is improving by means of this architecture ability to activate only modules required for application—by means of using a single 16-bit memory module, when necessary.

Table 9.3 presents a basic memory map with memory locations occupied by the ERA devices: ROM (2 banks) RAM (4 banks), USB, Ethernet, and UART interface. The 4 RAM modules are 1 Megabit each (64 k $\times$ 16 bit). The flash-based ROM modules are 128Mbit each (8 Mb $\times$ 16 bit).

The SRAM modules representing the units 8, 7, 5 and 6 are located in the highest part of the memory, followed by the ROM modules (units 9 and 10). The USB, Ethernet, Serial Ports, LED's and switches of the ERA board are mapped in the lower part of memory.

Fig. 9.5 Triplicated malfunction tolerant RAM, 1994 version

Table 9.3 ERA memory map

Memory range	Device details	Device
0700FFFFH –	ISSI-IS61WV6416BLL	SRAM logic module 2 (U7, U8)
07000000H		
0600FFFFH		
–	ISSI-IS61WV6416BLL	SRAM logic module 1 (U5, U6)
06000000H		
057FFFFFH		
– 05000000H	Sharp-LH28F128BFHT	ROM logic module 1 (U9, U10)
04000000H	FTDI-FT245BM	USB
0300FFFFH		
–	SMSC-LAN91c11i	Ethernet
03000000H		
02000002H		
–	RS232	UART Interface
02000000H		
01000000H	Normal	LEDs (1–2)
00000000H	KNITTER	Switch (3–4)

9.5 Conclusion for Chapter

1. In the development of reliable architectures there is a need for providing accurate testing and debugging of hardware and software. In this chapter, we first present the implementation details of hardware architecture primarily required for simulation.
2. We demonstrated ERA instruction execution flow and explained how it can achieve the decoding and execution of two instructions per single fetch and its implications on system compilers.
3. We also provide an overview of the hardware prototype and its memory mapping together with an architectural comparison to other relevant RISC and CISC architectures.
4. We introduced Minimal Instruction Set Architecture that simplifies the instruction decoder design and the overall system's reliability. ERA's ISA has 16 instructions, does not have a pipeline and provides only absolute memory addressing.
5. We argue how far from being a drawback, this simplicity is sufficient to perform safety–critical code improving efficiency reliability.
6. We provide the details of a testing and debugging methodology of a hardware prototype.
7. Finally we showed the features and implementation of an assembler, and a disassembler/simulator as a proof of concept of the architecture. These custom tools are useful, not only for testing and debugging of the hardware prototype, but for the system and application software.

Acknowledgements Dr Victor Castano and Prof Eugene Zouev were contributing in this chapter. Their help and devotion to our joint work make it possible to create useful software instruments for the processor IT-ACS LTD made.

Chapter 10
Conclusions for the Book so Far

Abstract This chapter presents an extended summary of the whole nine chapters. It explains the reason of each chapter in the context of the subject and highlight major findings.

Keywords Resilience · Reliability · Redundancy · Hardware · Fault tolerance · Interruption · Memory · Processor · Safety–critical systems

10.1 What We Have Done so Far

The main objective of our research was and is to find new ways to introduce new properties into computer architectures and understand the drawbacks of existing computer architectures.

At first, we reviewed the classic theories of reliability and fault tolerance. We were applying and elaborating reliability theory using as a key book by we are please to say Sandro Birolini Reliability Engineering (Birolini 2017) starting from 4th till 8th, the latest edition. His support and critics as well as navigation were and are priceless.

Additionally, to well-known statements such as:

1. The more components a system has the higher the probability of system failure and
2. The reliability of a system is often dominated by the reliability of its least reliable component.

We have discovered something way not obvious. Yes, Reliability theory is not applicable to analyze fault-tolerant systems directly—in reliability, we have to have a statistic of event to apply distributions and make some predictions. Fault-tolerant systems are all about one event that by all means must be excluded, recovered, tolerated.

But while we were making reliability models for various designs—in our case memories and processors we were able—provided all models were made with the same set of assumptions to make a pretty important observation.

© The Author(s), under exclusive license to Springer Nature Switzerland AG 2026
I. Schagaev, *Resilience of Computer Systems*,
https://doi.org/10.1007/978-3-032-00938-8_10

The more states, yes, degraded, reduced in quality of non-functional requirements a system or device has, the better reliability we can have.

Having two reliability models for two devices, we simply might divide matrixes of reliability (usually it is a set of equations around Markov model of a model of a system) for both and get a kind a "structural gain" or index—proof which devices have more reliability. In plain English.

— the more states system has the longer it can live.

At the same time, "the cost" of enrichment of reliability diagram for designed system might be prohibitive. The less resources we spent for bigger reliability matrix of the system the better.

Naturally, some of the keys to improve reliability would be simplicity of implementation and careful introduction and use of redundancies. We read redundancies as (Schagaev I, 90)—Yet another approach for redundancy classification.

There it was explained, and we follow in our books including this one three type of redundancies applicable to meet required NFR (non-functional requirements):

1. Informational redundancy
2. Structural redundancy
3. Time redundancy

Regarding principles of design—discussed and developed with another ETHZ guru: Prof Niklaus Wirth, Fig. 10.1 illustrates them and applicability.

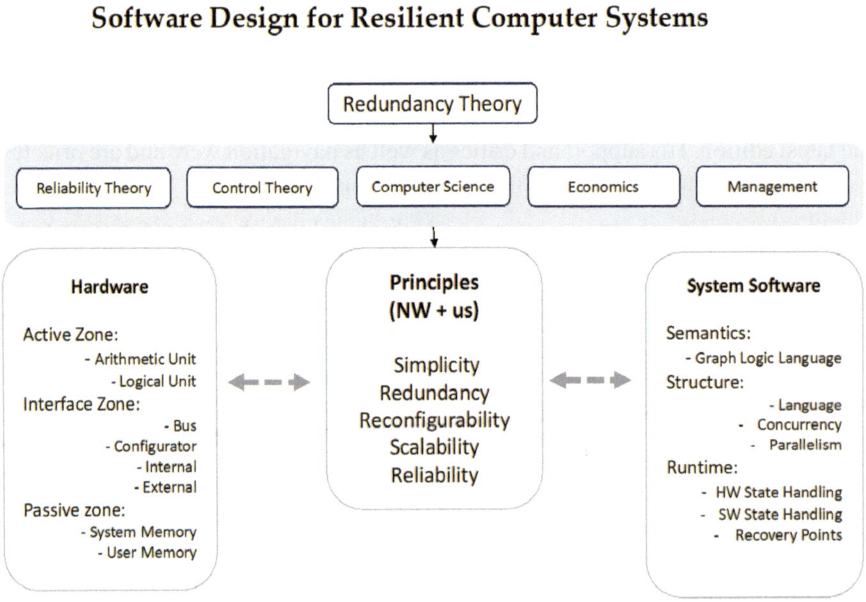

Fig. 10.1 Principles of resilient design

As far as we can see—the central box—*Principles* are applicable for both hardware design—left box—main part of this book, and right box—system software design—our recent book (Schagaev and Gutknecht 2024).

Now, about our resilience "facelifting". The terms Reliability, Fault-Tolerance and Dependability do not cover all the attributes and describe properties of safety–critical applications or, after being re-defined over the years, are ambiguous.

As a consequence, Chap. 2 provides an updated concept of resilience that suits better for safety–critical domain.

A resilient system, over a specified time interval, under specified environmental and operating conditions (performability), "must be ready" (in terms of availability) to perform its intended function (reliability), guaranteeing the absence of improper system alterations (integrity).

It must have the ability to conduct servicing and inspections (testability) so that, in case of failure, a quick recovery to a specified working condition is achieved (maintainability) to provide further functioning or discontinue system operation in a safe way (safety).

Furthermore, what is resilience should have:

1. A resilient system must have the ability to anticipate changes and evolve (evolvability)
2. Re-adjust itself during executing (adaptability)
3. Be able for accommodating changes by reconfiguring elements of the system if necessary (reconfiguration).

Since one of the keys to improve resilience was the careful choice of redundancy and the manner in which this should be applied, we decided to review the different types of redundancy and how such redundancy is translated into functional mechanisms to either avoid or tolerate faults.

We also introduced our own classification of fault-tolerant mechanisms based on the type of redundancy, we have employed and study their benefits and drawbacks. Fault avoidance techniques do not guarantee complete removal and protection from and of faults and present drawbacks such as cost, speed of operation and chip's area.

Therefore, fault tolerance mechanisms are critical and needed to make resilient safety–critical systems.

In order to select a specific set of redundancy types and effective techniques for their applications in the process of implementation of FT we should first define the different requirements of the particular application. Once the domain and requirements are defined we should select the techniques that suitable the requirements and the level at which the redundancy should be applied.

Further improvement of existing mechanisms of fault tolerance, is based on understanding where and why failures originate, what caused them, under which circumstances, in which contexts, and how often they happen After that rigorous analysis is required.

In Chap. 2, we introduced the concept of vicious cycle that explains our interpretation of the reasons behind the performance and reliability problems that jeopardize the application of Moore's Law.

We also reviewed the fault-failure lifecycle and defined the necessary concepts of fault, error, failure and catastrophic failure. Here our great thanks to Prof Jean Claude Laprie (since 80 s), his initiative and analysis of this kind of evolution. His efforts and personal material provided to us were simply priceless.

Since the majority of hardware faults in current electronics are induced by ionizing radiation we studied the damage mechanisms at the physical level, the sources of error and the micro- and macro- effects of such mechanisms. Physics of impact and technology simply enforces us to make Chap. 4 provides an extensive taxonomy of radiation effects describing their nature, type of degradation, susceptibility, fault rate trends and recoverability.

From the study of this taxonomy we conclude that as we moved to denser technologies at lower voltages, system SER (Single Event Rate) will continue to rise and in particular the contribution of SEU (Single Event Upset), SET (Single Event Transient), MBU (Multiple Bit Upset) and SEFI (Single Event Functional Interrupt) will increase.

We also conclude that current, known so-called "mitigation techniques" are not efficient when dealing with certain types of SEE and/or with the upcoming rates.

In Chap. 5, we explained how any fault tolerant system involves a Model of the System, a Model of Faults and a Model of Fault Tolerance. Consequently, we add value to such system by developing a comprehensive Fault Model suggesting methods for recognition and reaction against faults.

The following was discussed: fault manifestation, detectability, diagnosability and recoverability. The adequate solutions for diagnosis and recovery were proposed. We have introduced the principle of reconfiguration of the system and how this might be used for various purposes: performance-, reliability- and energy-wise gain, improving the efficiency of a system and its resilience.

In addition, we extended further our Generalized Algorithms of Fault Tolerance (GAFT) by providing the different states and actions required to achieve fault tolerance and therefore improve system resilience. This section is an extension of our previous works (Schagaev and Kirk 2025; Schagaev and Gutknecht 2024).

In Chap. 6, using know-how and conclusions acquired in the previous chapters we introduced a hybrid HW-SSW co-design approach of a resilient architecture with the ability to reconfigure, achieving various levels of dependability in different environments.

As part of the architecture, we first introduced the syndrome as a tool to handle new property of the system and analyzed it involvement in the process of reconfiguration—providing efficiency and gain in reliability, performance and power consumption.

Reactiveness is a core feature when system or object need to be operating in real-time safety–critical environment.

Thus, a special chapter (Chap. 8) was made to present an analysis of interruption schemes for resilient computers. Starting first from concept of interaction the specific features of interruptions for resilient system are discussed. Brief analysis of known architectures with interruptions were presented. The way how to achieve resilience and efficiency of new interruption system using our T-logic scheme is proposed in details.

From the top level of systems actor interactions, we elaborate what kind of resources and solutions are required to make interruption efficient.

We suggest that timers and other options of using various delays are not good practice to make interruption efficient.

Semantics of interactions in terms of available resources and synchronization is discussed going down to necessary schemes and algorithms of mutual exclusion.

Starting from classic conditions for synchronization and using our own models we design so-called fault-tolerant semaphores that, when implemented exclude process hang while within critical section of its execution.

We demonstrate how fault-tolerant semaphore can be implemented in a hardware of resilient system.

Shown that fault-tolerant semaphores increase resilience of the system as a whole.

Any theory needs practical confirmation to define applicability and efficiency limits.

Thus, our own design of embedded recoverable reduced instruction microprocessor as a part of the ERA (Evolving Reconfigurable Architecture) was presented. In this architecture functions of each hardware zone were defined regarding involvement in information transformation and processing: active, passive and interfacing. Our efforts in this segment are:

1. To implement fault tolerance the redundancy level needed to be as low as possible.
2. In the active zone, the instruction set and its implementation are reduced to the minimum; coprocessors, pipelining and floating-point units are removed which simplifies the processor design and reduces the complexity and fault rates.
3. Within ERA the checking schemes and re-execution of instruction mechanisms are explained presenting and how they reliability of a system can be improved.
4. Regarding the Interfacing zone, our so called T-Logic was introduced as basic element of reconfiguration and discuss its various configurations.
5. We introduced the syndrome and explained implementation details and how, in combination with a Memory Management Unit and a Reconfigurable Memory Scheme, it can act as a control center of three functions: fault monitoring, reconfigurability and recovery.
6. As part of the passive zone, the reconfigurable memory scheme can operate 25 memory configurations and support graceful degradation.
7. Markov model of reliability for ERA's configurable memory was presented, quantifying the probability of state transitions. Finally, in this segment of research a system software support for testing and reconfiguration we described. Shown that by combining this novel hardware architecture with the supportive system software, all key properties of performance, reliability and energy-wise functioning could be visibly improved.

In further chapters the implementation details of ERA's hardware prototype were provided. Having a software simulator of a hardware platform at hand is very useful to speed up software development and debugging of applications.

An accurate hardware simulator with graphical user front-end called Dissimera was developed and explained. Dissimera's main goal was not speed but to simulate reliably and accurately the basic core features of ERA with fully reproducible results.

The simulator is built based on principle of growing core: once core simulation is completed, it become possible to escalate simulator by adding new features. The development of such disassembler/simulator gives the following possibilities:

(1) Testing and locating errors of design of the soft core processor;
(2) Understanding the smallest details of the ERA functionality;
(3) Simulate of the current version of the processor and the FT version of the processor;
(4) Testing and debugging of errors in application and system software.

Finally, we introduced a testing framework that in combination with Dissimera's, with ERA's assembler and with commercial hardware simulators can properly test and debug not only the ERA's hardware prototype but ERA's application and system software.

10.2 Next Steps

Arithmetic and logic units are both implemented through the use of logic components. It is known that an arithmetic instruction can be translated into several logic operations. Applying this principle, if an arithmetic unit is suspected of not being able to provide correct service, arithmetic instructions can be translated into logic ones that can be executed by the logic unit of the processor.

Further research could be done on determining if logic operations can be translated into a set of arithmetic ones and how it can this be implemented. What would be the complexity of such translation? Performance would be affected (graceful degradation), but this technique would allow a running program to finish before recovery or fail-safe restart takes place.

The impact of the size of the register file on overall performance of processor is also a question of further research, as in 6.2.1.

Regarding Dissimera, although basic functionality has been achieved, the implementation of Dissimera is still a work in progress:

1. The design is completed and the user interface is fully defined.
2. Assembling of pseudo code using ERA assembler-preparator (100% completed).
3. Disassembling of binary into human readable code (assembly code) (100% completed).
4. The simulator is capable of parsing the binary file resulting from the previous step and is then capable of classifying data and instructions (100% completed).
5. Simulation of main memory, register file, program counter and instruction registry is almost completed (90%).

In future revisions, we have the intention to include support for: syndrome, extra memory configurations including 16-bit memory configurations and fault-injection. Also, we are:

1. Very interested in find ways to exploit more the functionality that the syndrome can provide.
2. We have no doubts that for safety–critical missions such as embedded systems in satellites or space further research is needed.
3. Would like to pursue more research in dependency matrix mapping of symptoms and failure modes, dealing with traces of consequences taking into account size of the damage domain and time duration of recovery actions.
4. Would like to apply the context sensing (e.g. altitude, latitude, temperature, dynamic events such as solar flares and weather forecasting) and experience to system software in combination with the syndrome to be able to see and prove the level of resilience of proposed approaches and our architecture, maybe with adequate adjusting.

References

Birolini A (2017) A reliability engineering. Springer Berlin Heidelberg. 978-3-662-54208-8

Schagaev I, Gutknecht J (2024) Software design for resilient computer systems. Springer Nature Switzerland. https://doi.org/10.1007/978-3-031-55139-0. ISBN 978-3-031-55138-3, ISBN 978-3-031-55139-0

Schagaev I, Kirk B (2024) Active system control, design of system resilience. Springer Nature. https://doi.org/10.1007/978-3-031-77181-1. ISBN 978-3-031-77180-4

Schagaev I (1990) Yet another approach to classification of redundancy. IMEKO symposium CIM. Helsinki, Finland, pp 117–124

Chapter 11
Vision on Evolving System Future

Abstract As a future vision, we present here an approach to design of evolving computer system. We summarize achievements of others and highlight the drawbacks of existing systems, explaining limitations of hardware and software co-existence. New design principles, models of representation for algorithms and architecture are introduced and described with an explanation regarding how to achieve better parallelism and reduce concurrency. Structure of required system software and hardware for reconfigurable adaptable system is presented and explained. Prototype architecture of evolving system of computer is discussed in terms of gain of performance, reliability and power saving. Comparison with Berkley approach to future computer is given. This is an extended version of our papers and tutorials presented at World-Comp 2014, July 19–24 Las Vegas, USA. Simon Monkman, from Bath and Prof Luc Blaeser, from Zurich kindly contributed figures and Composita Language examples in this part of work.

Keywords Redundancy · Reconfigurability · Recoverability · Performance- · reliability- · and energy-wise systems · System software · Hardware design · evolving systems · coverage

11.1 Fundamental Problem

A sequential way of thinking developed by the human race through evolution was constructed on a framework of languages and accompanied grammars Williams (2003).

Further, differences in dictionaries, alphabets and grammatical rules were visible, but not critical Schagaev et al. (2020). Indeed, we speak, write and listen to each other sequentially—word after word, phase after phrase—thereby creating a *one-dimensional* sequence of information.

With the appearance of technological support for logical and arithmetic calculations, computers implemented a system of processing which mimicked our habit of

© The Author(s), under exclusive license to Springer Nature Switzerland AG 2026
I. Schagaev, *Resilience of Computer Systems*,
https://doi.org/10.1007/978-3-032-00938-8_11

"sequentialism". Calculations were implemented on developed hardware assisting sequential way of thinking.

Modern hardware technology squeezes logic elements down to the nano-sized with changed hardware structure, mapping electronics into a matrix of interconnected blocks—in other words, hardware topology became *two-dimensional*.

There has been a recent attempt to make 3D chips of memory and processors—a simple Internet search indicates (40 million hits on the subject) that since 2007, IBM, Intel, Samsung, Toshiba and others are progressing towards the production of three-dimensional chips. There are declarations that Moore's law will finally be overcome and with the corresponding boost in performance for future computers being substantial.

It is worth mentioning here that "Moore's law" actually is not a law of nature, or physics, thus a discussion of any limitations to "Moore's law" is in some sense irrelevant in the first place.

This is a subject for research regarding the impact of mass media on technological developments; though social science researchers might find it interesting.

There is an unsettling feeling, though, that the professed goals will not be achieved. This feeling is based on a simple model: imagine a rope, of substantial length, which we need to put into a square box of much smaller size.

There is no doubt that this process will take time and the boxed rope will not be as convenient to use as straight one. Now imagine that the box became three dimensional; it is debatable whether our ability to use rope efficiently will grow at all…

Adding the need to accept dynamic changes to some segments of rope—updating, or deleting them and then reassembling the rope as a whole, typical functions of run-time system—can reduce any optimism about the impact of 3D to next-to-none. "Rope" here means, of course, a generalization of a program with code and required data.

Figure 11.1 illustrates some of the problems we are facing in computer design. It is clear that the marriage of one-dimensional programs with two-, or three-dimensional hardware may not be a successful one. It is also clear that the inclusion of dynamic support for this collaboration will cause even more complications.

11.2 Known Solutions (What We Have…)

Computer science discovered parallel computing in the late 40 s to early 50 s of the last century: Goldstine and Goldstine (1946), Everett et al. (1947), Whirlwind (1952), Hofstra University (1999), Holland (1960), Newell (1960), Schwartz (1961), Slotnik et al. (1962), Squaire and Palais (1963), Gosden (1966).

The real boom started just a few years ago. An absolutely stunning level of technology and hardware density was recently achieved Nair (2002), LaPedus (2003) with processor frequencies up to 4.7 GHz and beyond IBM (2007).

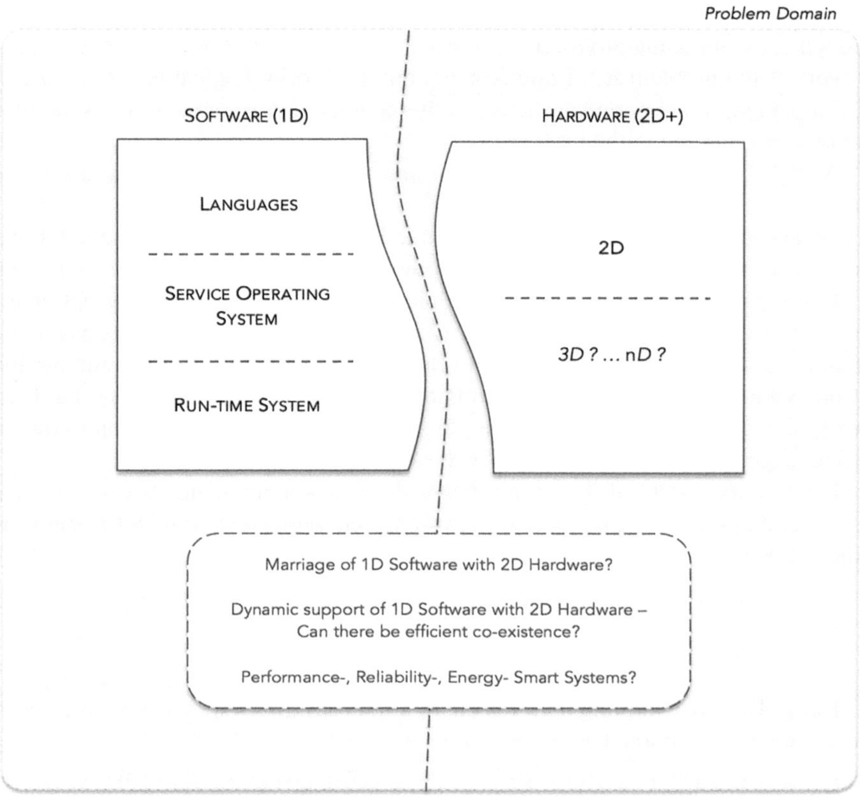

Fig. 11.1 Computer design problems

Surprisingly, the main questions in computer design and technology are still the same as half a century ago. Specifically, the most challenging goal is the design and development of a system that properly uses the relative dependencies of performance, power consumption, cost, reliability, adaptability and flexibility.

None of these dependencies are studied *together* except Monkman and Schagaev (2013). Nobody even set a goal to make a system that will be efficient and flexible along the lines previously mentioned.

It is clear that we need to describe the relationship between performance, reliability and energy-efficiency of architecture, ability to "trade-off" these properties through design, and observing how these properties change over the life cycle. It was only initially described by us in Monkman and Schagaev (2013).

The theory and development of a system with the required properties that considers those properties as dynamic processes, as well as their interdependencies and modifications (evolution), is in its infancy.

What we have to do now in ICT does not look encouraging: in system design, we have failed to achieve even doubled performance by applying a doubled frequency and quadruple energy, we have failed to achieve any visible system performance growth using the same programs.

Part of the problem is, of course that computer technological progress was not accompanied, or supported sufficiently, by theoretical development, i.e. something serious was missing in the first place.

Applying our mantra: "*Any theory values by power of prediction and applicability.*"

We accept that computer science has failed to lead the ICT world and has closed down discussion of technological developments.

In system software, the situation is confusing and again, not overly optimistic: an overcrowded family of languages (2500 members), Kinnersley (2009), with slightly fewer run-time systems (500+) Berka (2009) shows confusion and absence of breakthrough developments in terms of efficient system software and hardware designs, and coexistence. The market domination of one or two operating systems, or languages has nothing to do with a "best player wins" rule.

For example, state-of-the-art hardware illustrates a complete loss of direction; Intel's attempt for an eighty-core processor system has ended up with the statement Ames (2007):

> ...Despite using such an efficient grid, the researchers found they could actually hurt performance by adding too many cores. Performance scaled up directly from two cores to four, eight and 16, performance began to drop with 32 and 64 cores...

Examples prove nothing, but it is clear: a new design strategy is required for the next generation of computer architectures. It should start right from the design phase and be pursued through the whole life cycle of the system and applications.

Any new design strategy should have the goal that the system evolves with support from both hardware and system software.

This feature of "evolution" should be available on demand by user applications and operational requirements. It is also clear that the next generation of computer systems has to address all of the three prerequisites for evolution: Performance-, Reliability- and Energy-smart functioning.

They should be pursued from the first phases of system design, enabling an efficient trade between *P*, *R*, and *E*, if user or environment requires.

11.3 Attempts to Evolve

Our previous attempts to revise existing designs and find balanced developments of ICT hardware and software were presented in Schagaev and Gutknecht (2024), Kaegi (2013), Plyaskota and Schagaev (1995), Monkman and Schagaev (2013). This work summarizes our earlier concepts and developments. At first, we propose a revision of the area of computer designs and development by introducing a new system paradigm. Any paradigm has to be useful—see our mantra above. Thus, we have to develop supporting theoretical models and prototype system in both software and hardware. Several holistic principles we have been using, are as follows:

- Simplicity and redundancy;
- Reconfiguration and scalability;
- Reliability and fault tolerance;
- Energy-wise design.

These principles are explained to some extent in Table 11.1.

As described in Kaegi (2013), the declared principles have to be pursued by applying redundancy in the form of *information, structure* and *time.*

The whole process has to be revised from concept through development and the construction of algorithms, the further joint design of hardware and system software and implementation and maintenance, each time pursuing a goal of simplicity and introducing essential redundancy.

This would define the ability of future systems to reconfigure themselves for the purposes of Performance-, Reliability- and Energy-smart operation. It means also that new systems have to provide an efficient process for "trading off" reliability, performance and power consumption.

Reconfigurability might be implemented efficiently only when hardware designs are supported by system software solutions.

Semantics and the structures of computer hardware vary. The area where information is transformed (we call it the Active zone) is complex and largely irregular. The Passive zone of hardware—where information is stored—has a regular structure

Table 11.1 Holistic principles of computer system design

Simplicity	Complexity is difficult to implement and handle efficiently. To reduce complexity, 'bells and whistles' in the architecture are excluded, together with compatibility with main market players, or conventions (pipelines, caches, TLB, etc.)
Redundancy	The deliberate introduction of hardware and system software redundancy and monitoring schemes provides the required level of reconfigurability to reach performance and reliability goals. Redundancy has to implement three processes: checking, recovery preparation and recovery
Reconfigurability	Reconfigurability serves three main purposes: performance, reliability and power awareness. Handling reconfigurability should be done using Graph Logic Model and run-time support. Active support for reconfiguration during design and at run-time by hardware and system software resources is required for scaling the system up to 100 + processors
Reliability	Minimum hardware redundancy for design of highly reliable main components. Redundancy of system software and hardware maximizes both: tolerance to malfunctions and permanent faults. System software, therefore, supports hardware fault tolerance
Resource-awareness	Mission critical systems have significant limitations in terms of hardware resources and power consumption constraints, (e.g. battery life). Thus, for wise resource use, reconfigurability must be introduced

and the highest density of transistors. An Interfacing zone acts as a bridge between the two and serves to maximize the speed of data exchange.

Table 11.2 describes how these principles can be supported and followed during the development of new hardware architectures (HW) (grey boxes) and system software (blue boxes) (SSW). When used in combination, it might allow the next generation of computers to be free from existing limitations.

In terms of information processing, computer architectures can be divided into three areas: the processing area—further called the Active zone, the storage area—called the Passive zone, and the Interfacing zone. All three zones have to be reconfigurable for their own purposes and the needs of other zones.

Dependencies between zones define the level of reconfigurability and flexibility of the hardware and system software parts of the system architecture. Note here also, that *P*, *R*, and *E* are no longer system requirements, as they become *modifiable* during life cycle: from the first phases of system design Plyaskota and Schagaev (1995) down to maintenance.

Therefore, the requirements of evolving systems need themselves to be considered as processes. *P*, *R*, *E* dictates that system software has to be modified, including language, run-time and service operating systems. Figure 11.2 qualitatively shows that system Performance- (*P*), Reliability- (*R*) and Energy- (*E*) smart design might be implemented by means of various methods (circles) and to achieve any development, we should choose only not-mutually exclusive solutions.

It is worth mentioning that the properties of system software that are required and useful at program preparation time, are not required at all during program run. Obviously, they should be "stripped off" program code and instead, a "fine tuning" of the code applied to fit existing hardware.

The communication, or interaction between agents, processes, communications, monitors (use any other name from modern avalanche of terms), should also be implemented by taking into account *P*, *R*, and *E* requirements, as well.

One of the immediate observations that can be made from Fig. 11.2, regarding performance, is that: the parallelization of a program should be prepared and supported as much as the program structure enables, or even more, while concurrency within the program and during execution should reduced to a minimum, or excluded.

It is beyond the scope of one chapter to describe all possible developments of system software and hardware for *P-R-E* systems. Below, we present just one approach, attempting the redevelopment of system software for performance improvement.

Table 11.2 Properties versus components of the system

	Simplicity	Redundancy	Reconfigurability
Active processors	Only essential for instruction execution hardware should be developed, duplicated instruction "for convenience" or compatibility should be excluded; no not necessary back-size actions of instruction execution, such as signs, flags etc	Extra elements and schemes should be added either for improvement of reliability or energy-wise functioning	Designed processing element should be internally reconfigurable (not used hardware is disabled) and support of reconfigurability of the system
Interactive (internal and external interfaces)	Minimum hardware with when possible state-less designs should be used, using asynchronous logic as much as possible	Designed interconnections must be identical in terms of functioning, making maximum reconfigurability possible, able to be used as direct access in and onward	Control of physical and virtual addresses
Passive (memories)	Reduced complexity of address schemes (internal and external)	Virtual memory (of the same type)	Ability to switch datas and program areas, hierarchy of recovery points
Language	At language level -cancelling of complex data structures; limitation and simplification of program descriptions and code structure	Generation of more than one code to run a program for various requirements: *P.R.E* Automatic hierarchy formation of recovery point (also for reliability of operation)	Semantic modification of program (graph-logic and control data predicate models, see further) to reduce dependencies between code segments and data; support at the language and operating systems-service operating system and run-time system), processes

(continued)

Table 11.2 (continued)

	Simplicity	Redundancy	Reconfigurability
Monitor of processes	Simplified support of process synchronization	Reducing amount of supportive data in run-time system for program monitoring Introducing of reserved data blocks to run hardware with deficiency (for reliabilitypurpose of view)	Implementation and control of software during run-time and hardware-making "integrity map" possible
Reliability support	Recovery points of a program, modules and tasks—to implement hot swap (reconfigurability) when needed. Saving only essential data for concrete execution of a program		

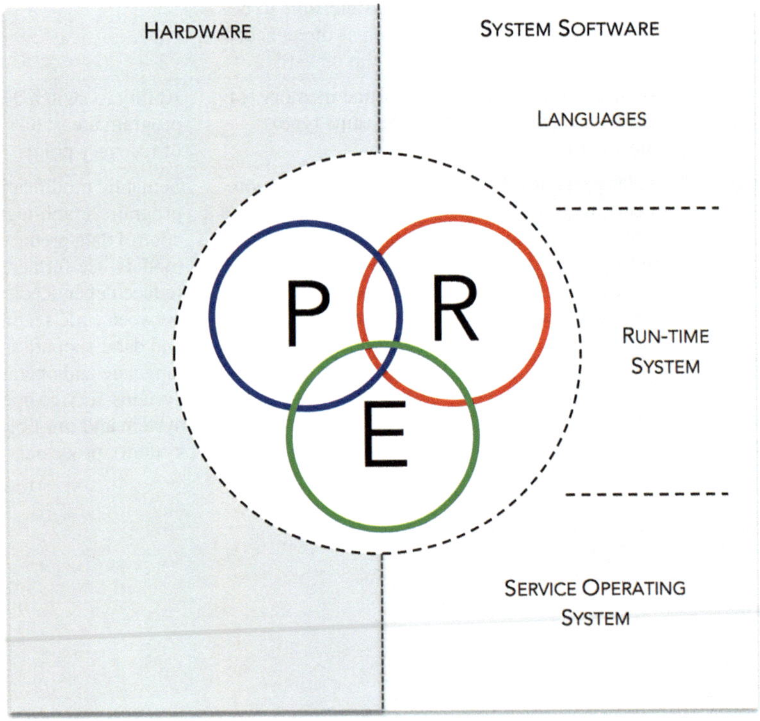

Fig. 11.2 System properties reflection at system software

11.4 Proposed Approach (What We Need and Why We Need This)

At first, we should choose the nearest, "rigorous" language and run-time system and define a sequence of steps that lead to the modification of the system software with the aim to improve reconfigurability and other requirements shown in the Tables above.

To make reconfigurability work, we describe program and hardware using meta-models. These meta-models are independent from the technological aspects, but describe the structural properties of the algorithm and programs: Control Data Predicate (CDP) and Graph Logic Model (GLM).

The languages of structural programming, modular programming Turski and Wasserman (1978), Wirth (1983) and object-oriented programming Wirth (1989) Wirth (1992) were invented to provide the programmer a higher level of abstraction.

There is no doubt that the separation of concerns increases visibility and clarity and indeed, eases the process of programming. These properties, unfortunately, are not necessary and even reduce performance during execution, when hardware load and power consumption are crucial.

In terms of *P*, *R*, *E*, the properties of the system all attempt to deal with parallelism and concurrency of programs and their execution at the same time. They are still far from providing substantial gain, in spite of extreme efforts Flynn (1972), Fisher (1983). Dijkstra (1965) has published an extension of the Dekker algorithm for a solution of the concurrent execution of multiple processes and it is this solution, in one form or another, that is used within languages and hardware (instructions like "test and set", (IBM360).

The problem is the automatic extraction of the program parts that are independently executable; parallelism is still a "work in progress" and attracts the attention of researchers on a regular basis, all around the implementations of mutual exclusion and synchronization Lamport (1983, 1985), Gutknecht (2006).

In discussions of performance gain system from software parallelism vast majority of publications are oriented on Amdahl's law. However, Amdahl's simplification of the efficiency of parallelism by cutting—see our afore-mentioned rope in n chunks—does not reflect the fact that chunked segments *should be delivered* into the processing areas, and the results *subsequent to* processing *should be collected back*, usually into one block of data. Read: in Amdahl ratio denominator grows, *reducing* gain from parallelism even further.

Thus, the roles and impacts of language support of parallelization, service operating system and run-time system were largely ignored. Therefore, predictions of Amdahl about the gain achievable by parallelization were, to put it politely, extremely optimistic.

Works and publications Nair (2002), Lapedus (2003), Hennesy (2025), Bryant and O'Hallaron (2002) highlighted problems of massive parallelism in future hardware designs, showing that the way we approach them, either limit seriously, or makes impossible to cope with, the requirements of configurability and scalability:

What is good for one purpose is unacceptable for another.

Due to architectural and technological limitations, known hardware designs are not flexible enough in the mapping of algorithms to the hardware, and above all, do not support changing of software structures or hardware configurations "on the fly".

Commercial systems (Windows, Linux, etc.) have shown that attempts in the effective use of hardware features have largely been a failure: the chip frequency now goes beyond 5 GHz, but hardware design and existing software reduce useful system performance down to the 100 s of MHz range. To cope with this, future architectures must be re-designed with a justification of concepts and arguments from the user level of abstractions, down to the hardware level of representation, aiming at minimizing the degradation of any required characteristics.

Technologically, new hardware chips have seemingly astronomical numbers of logic gates and high physical density. This gives substantial advantage for programming multiple-data problems, where it is possible to use matrix algebra, or multiple data tasks with simple instructions. In turn, the density of hardware elements has reached a technological limit of thermal density.

As already mentioned, accepting the principle of separation of concerns, the programming language has to be hardware independent, as in this way, the logic and complexity of an algorithm will be visible and separated from hardware implementation. From another point of view, we should improve efficiency of hardware and software work.

So, the question remains:

How do we achieve the PRE-smartness of a system by modifying algorithms and implementation?

Certainly, the compilation process of a program should be "fine tuned" for execution on the specific hardware. Additionally, though, reconfiguring hardware before program execution, also reduces hardware overheads.

Examples prove nothing, unfortunately. In much more general terms, program tuning for hardware Performance, or Reliability, or Energy-smart operation is becoming some kind of reversed programming, where hardware, not the user (programmer) takes the dominant role.

Thus, we might explicitly seek to separate "what is good for the user from what is good for the system and hardware" and, where possible, spread these requirements through the life cycle of software and system development, as Table 11.3 illustrates.

The separation of concerns and goals of PRE-smartness requires supportive models of algorithmic applications from the system software point of view (language and run time systems) and at the same time, improve hardware efficiency.

Two models we propose for this are called Control Data Predicate (CDP) and Graph Logic Model (GLM).

Table 11.3 Phases of software development

User oriented phases (UW)			
Concept	Design	Compilation	Implementation
System software and hardware oriented phases			
Analysis of the program through graph logic model to find intrinsic parallelism in control, data and predicate dependencies	Recompilation of the program into parallel form with introduction of hardware configuration and reconfiguration features into the program, as well as concurrency management—formation of an execution string		Dual string execution (New RT data structure that unites arrays and records)

11.5 Supportive Models

11.5.1 Control-Data-Predicate (CDP) Model

Within CDP model each operator in a program can be defined in terms of modifications of three connected graphs: Control (C), Data (D) and Predicates (P), illustrated in Fig. 11.3.

The Control graph presents a sequence of instructions for assembler or processor, or operators for language. Data dependencies related to each operator and data that might be modified along program execution are illustrated by the Data graph. Finally, a graph of Predicates defines the change of conditions inside hardware and programs to enable branching. Thus, every step of program execution is described and defined by a change in three graphs.

While the program operators define a number of elements (nodes) in each graph, the volume of the resulting predicates, state registers, processor and Program Status Words (PSW), physically represented in the hardware by so-called processor status registers, contributes to the "width" of each layer of these graphs.

To support and exploit parallelism, we should simplify all three graphs by separation of Data and Control, making them as independent as possible.

To support simplicity, a graph of Predicates should be simplified and fragmented—in other words, we propose that Predicates should be processed when they are needed for decision-making and not stored in the hardware.

To support reliability and fault tolerance, the detection of possible changes to program state caused by hardware deviation is required, together with the development of recovery schemes of hardware and software Kaegi and Schagaev (2013). Therefore, any complex operator and hardware instruction respectively jeopardize the possibility of generating "snapshots" of the previous hardware states, as well as the flexibility of program segmentation, allocation and reconfiguration.

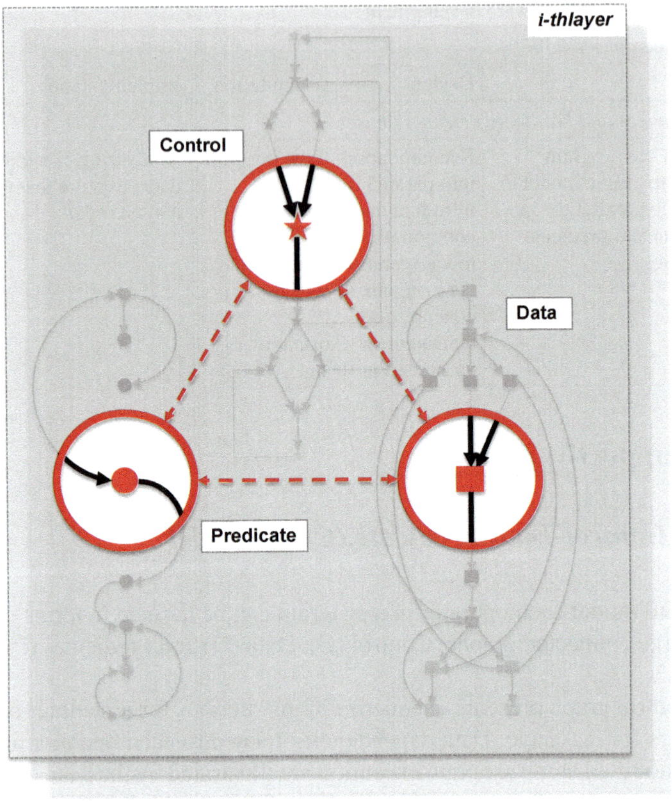

Fig. 11.3 Control data predicate (CPD) model

In the vast majority of architectures, as well as languages (their compilers) and run-time systems, the functions of data access and data processing are mixed—tightly coupled—and hardware state modification due to program execution is not controlled explicitly. This makes condition change latent and execution of a program unjustifiably complex.

In a typical processor, such as ARM, Intel and SPARC, the Arithmetic and Logic Unit (ALU), or even several of them, as well as shifters, registers, internal cache, special registers and pipeline sequencers, etc., are active during the execution of each instruction, sometimes with several data passes within a single instruction.

Thus, the complexity of hardware handling becomes enormous: 75% of the die size is occupied by translation look-ahead buffers, caches, synchronization logic and pipelining. However, none of these overheads are required from the point of view of the programming language and program operators.

Any condition of the hardware related to an operator, or instruction representation (the three nodes within the same layer in Fig. 11.3) requires checking. This makes parallelism, or reliability much harder to achieve. The reasoning is different, though:

for parallelism, hardware should be designed as "flat as possible", but reliability demands the limitation of fault propagation through hardware schemes.

The complexity of a system and the implementation cost of parallelization, or fault tolerance, are directly related to the amount of the resulting modifications of the hardware and program states.

Our CDP model shows that when P is only used for the selection of the program flow, a special operator and instruction can be defined to generate the current value of P and store the result in a register, making system reconfigurability easier to achieve.

To summarize, for the implementation of parallelization at the level of the instruction set, the design objectives will be:

1. Implementation of "as simple as possible" logic to form predicates;
2. Mapping the language operators as close as possible to the processor's modified instruction set;
3. Reducing the size of the state space that needs to be saved before the execution of each instruction.

This brief analysis of CDP shows that clarifying the interaction of language and hardware can save us from "improvements" that drive mutual loss. Studying the CDP scheme of a program also allows potential program parallelism to be checked in all three graph dependencies. However, this is not the whole picture.

11.5.2 Graph Logic Model (GLM)

As a further improvement in parallelization and at the same time concurrency reduction, we need to handle the actions that were initiated in parallel, but eventually end up in conflict when one or more resources are accessed.

To describe this, we introduce the Graph Logic Model (GLM), Fig. 11.4. Every meta-program structure might be described using GLM. GLM provides a scheme to redevelop existing programs into their maximum parallel and minimum concurrent forms, limited only by available hardware resources.

An indicative example of how the GLM works, using a program control graph, is presented in Fig. 11.4.

Note that a GLM might be applied for any of graph of the CDP model. GLM uses logical operators from the set {AND, OR, XOR} for every program, or hardware scheme that it describes. These operators are allocated for the input and output of each vertex. A vertex might be an operator, instruction, or state. Vertex **a** of Fig. 11.4 may be described thus:

$$\mathbf{a} : \text{OR} - (\alpha\mathbf{b}, \gamma\mathbf{d}), \text{AND} + (\beta\mathbf{b}, \delta\mathbf{c}) \tag{11.1}$$

where "−" stands for every logical operator of an output link and "+" for every input link, while α, γ, β, and δ are weights, or priorities assigned for the link.

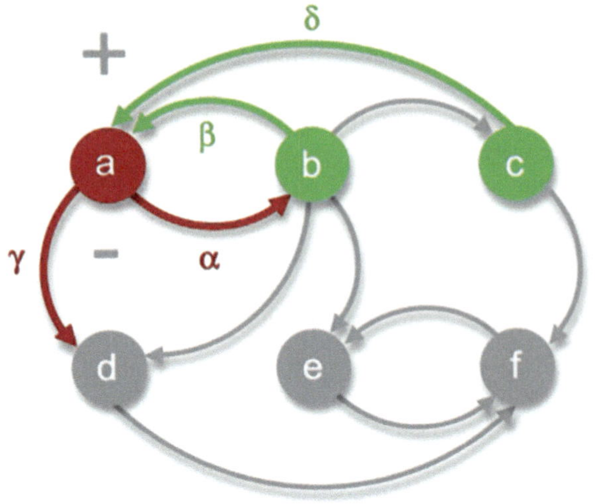

Fig. 11.4 Graph logic model

Until now, research in parallelism has been mostly targeted at finding parallel branches of programs and independent data elements.

However, expecting pure parallelism is hardly feasible; what is initiated as parallel segments ends up ultimately in concurrent mode, competing for a resource such as a socket, printer, or data concentrator, etc.

The rare exception, such as graphic processors with high numbers of SIMD-like processors, just proves the rule. A simple notation, similar to (1), can describe program structures and hardware structures consistently in terms of co-existing concurrency and parallelism.

GLM *explicitly* separates parallel and concurrent elements in the system description by introducing logic operators into the program graph for incoming and outgoing ends of edges.

The application of the logic operator XOR (exclusive OR) on an input, or output of an edge, defines ALL possible concurrencies in the program graphs. In turn, all possible parallelism in the Control graph is defined by finding all outgoing, or incoming edges explicitly described by the AND operator. The same approach might be applied for the Data and Predicate dependency graphs of the CDP model.

Moving forward, obvious questions of PRE-smartness arise: "Can we prove that the hardware representation for the chosen algorithm is parallelized in its maximum possible way?" "How does the current hardware-software solution correspond with a limitation of available resources?".

The answer to these questions is a part of the design challenge for any evolving system. Having a correct program, CDP and GLM can then be applied to extract the parallel segments and data paths and help in reducing concurrency.

This reversed programming gives us a chance to play with the software and monitor system software redundancy, (deliberately introduced at the recompilation phase) and hardware redundancy (introduced at the design phase) on the fly.

It also enables to use reconfigurability of the system for Performance, Reliability, or Energy-smart functioning when it is necessary.

A new language is needed to naturally express algorithms in a form that supports program recompilation into a redundant form fitting with the "hardware view". This does not mean that other programming paradigms, such as structured, or OO is no longer useful. It means, though, that they have their limits and mostly serve user convenience and interfacing.

A new language is needed to naturally express algorithms in a form that supports program recompilation into a redundant form fitting with the "hardware view". This does not mean that other programming paradigms, such as structured, or OO are no longer useful. It means, though, that they have their limits and mostly serve user convenience and interfacing.

The newly proposed paradigm serves to improve performance of the end product. This is achieved with the design of a new reconfigurable architecture, a system software and the dynamic transformation of existing software into a form that is efficiently executable by adaptive hardware: Evolving Reconfigurable Architecture (ERA), or more generally, Evolving System (ES).

Interestingly, the algorithm of parallelization might be applied for all three dependencies graphs—Control, Data and Predicate—and, therefore, deduce the maximal parallel form. The availability of hardware serves as the termination condition for this algorithm.

GLM explicitly separates parallel and concurrent elements in the system description by introducing logic operators into the program graph for incoming and outgoing ends of edges.

The application of the logic operator XOR (exclusive OR) on an input, or output of an edge, defines ALL possible concurrencies in the program graphs. In turn, all possible parallelism in the Control graph is defined by finding all outgoing, or incoming edges explicitly described by the AND operator. The same approach might be applied for the Data and Predicate dependency graphs of the CDP model.

Moving forward, obvious questions of PRE-smartness arise: "Can we prove that the hardware representation for the chosen algorithm is parallelized in its maximum possible way?" "How does the current hardware-software solution correspond with a limitation of available resources?".

The answer to these questions is a part of the design challenge for any evolving system. Having a correct program, CDP and GLM can then be applied to extract the parallel segments and data paths and help in reducing concurrency.

This reversed programming gives us a chance to play with the software and monitor system software redundancy, (deliberately introduced at the recompilation phase) and hardware redundancy (introduced at the design phase) on the fly. It also enables to use reconfigurability of the system for Performance, Reliability, or Energy-smart functioning when it is necessary.

Figure 11.5 introduces the use of CDP and GLM for program parallelization. There is no doubt that we have to start from correct sequential program.

1. Debug sequential algorithm (SA)
2. Create a model of the SA as triple <C, D, P> where C, D, P are graphs of control, data and predicate dependencies of SA
3. Apply the graph logic model to modify C, D
4. Where possible substitute "strong" logic operators by "weak" ones. An "IF" might be substituted by XOR (to start) and OR, or AND or NOP, leaving an option to parallelize segments of the algorithm
5. Check all introduced XOR in sequence for potential concurrency in the algorithm – for the OS

Fig. 11.5 Sequence of steps to make sequential algorithm parallel

The proposed sequence clearly separates properties required at the program-writing phase from the properties required during execution, as the latter are dependent on the hardware resources that are available and may change during execution. In this way, hardware features (that might be dynamically adapted during run-time) become represented in the program logic.

11.6 Coverage—Generalized Model

While achieving the system performance via parallelism we hope became clearer if one apply graph-logic model and control data predicate model as well, an idea of system level of reliability needs some work. This short section is just about that.

In principle, an Information from the object might be used for reliability of any system when model information about the object can aggregate it Reflecting (R) it in a state domain Ω.

This model is described by transformation T of the state domain

$$\Omega : T(\Omega) \rightarrow \Omega^{+}$$

Note that the transformation T is an internal feature of the model, not of the object itself. So, the model of information about an object or process can be based on the analysis of the changes of Ω. Growth of Ω can be achieved by:

1. Growth of "performance" R, while the reflection procedure has sufficient power to process more info from the existing flow of information in a timely way;

2. Growth of time t to form Ω using the same R, while one can use more time to provide more precise prognosis i.e. the results are not required urgently;
3. Auto-generation $A(\Omega) \rightarrow \Omega+$, $\Omega+ \supset \Omega$ where a new state of domain $\Omega+$ is generated, sometimes deliberately based on the existing Ω to achieve some additional meaning or benefit.

With active system control and, for example, reliability—availability or real-time safety in mind the third variant of Ω growth seems questionable as it does not involve receiving any new information from the object for its translation into information for the model and further decision making. This is a well-known phenomenon in any system where the human factor is involved, for example bureaucracy tends to create more bureaucracy but not efficiency…

The two other options are much more attractive: growth of time t having the same R increases set Ω up to $\Omega+$, and if $\Omega+ \supset \Omega$, it means that new information about the object is presented in the model.

On the other hand, growth of performance of R so that $R \rightarrow R+$ changes the pair $\{R, \Omega\}$ into the pair $\{R+, \Omega+\}$, where $\Omega+$ can also be substantially bigger than Ω.

Some correlation measure between the "physical" state of the object (say, power plant, a car or an aircraft) and its level of control and safety needs to be established. Let us assume here that it is a probability of operation success.

Now process or an object improvement is concerned with the analysis of $\{R, \Omega, T(\Omega), P\}$, where P is the probability of successful operation completion, and all the rest are presented above. Other probabilistic measures were presented in our book Active System Control (Schagaev and Kirk 2025).

Practically useful directions of active system control in terms of reliability, availability and real-time safety regarding theory development are:

- To change Ω so that $\Omega \rightarrow \Omega+$ to increase P up to required $P+ \mid P+ > P$;
- To change $T(\Omega)$ to provide growth Ω up $\Omega+$ and P up to $P+$;
- To vary Ω and T so, to achieve maximum of $P \mid P$: $\Phi(\Omega, T(\Omega)) \rightarrow P\max$.

The symbol Φ here represents a subset of the states in Ω that describe a faulty or strange behavior of an object or process. Between them Φ and Ω define a probability function of successful operation. The transformation of existing space Ω by means of T to $T(\Omega)$ describes the role of active control and safety model and can be used for evaluation of achievable control, safety and the efficiency of the object or system design and model itself.

The connection of these tasks to real time and aggregated previously data is determined by the size of Ω, the performance of algorithms and hardware ($T(\Omega)$) and, from a different viewpoint, the statistics based on accumulated data.

So, in terms of information, the task of analysis of process success can be described by:

$$Iu \rightarrow Ru(Iu) \rightarrow I\,m \rightarrow R\,m(I\,m) \rightarrow \Omega \rightarrow T(\Omega) \rightarrow P$$

where:

1. Iu, Im are real information about the object and model information respectively, and
2. Ru, Rm are the reflections (transformations) of the features of the object into features of the model, and

$T(\Omega)$ is the transformation of the set of states in the model for further analysis of the probability of successful flights.

Growth of Ω into $\Omega+$ might infer an indirect increase in information about the object, but only within the limits imposed by physical laws.

The number of states connected with fails is presented in the model as Φ, and $\Phi \subset \Omega$ and $\Omega \subset \Omega+$. This means also, $\Phi/\Omega > \Phi/\Omega+$ and, using the model with a larger set of states, it should be possible, at least in principle, a reduction of risk of an accident or fault, as—we are able to detect adjacent the border states.

The growth in the number of states is determined by $(\Psi/\Phi)/\Omega+$ where Ψ goes from $\Psi/\Omega+ = \Phi/\Omega$.

Figure 11.6 illustrates the relationship of the set of these states.

When $\Phi \subset \Psi$, then there is an option for prognosis of system state and success will be more likely when the "smoother" and "regular" set of states Φ is enclosed by set Ψ. It is worth pointing out again that Φ represents the set of accident (risk) states and fortunately it will not grow when $\Omega+$ grows. Figure 11.6 was made with an assumption about coverage (our knowledge about the system or model) is a constant that represents graphically a proportion WWK/OMEGA:

$$C = \frac{\mathrm{WWK}}{\Omega}$$

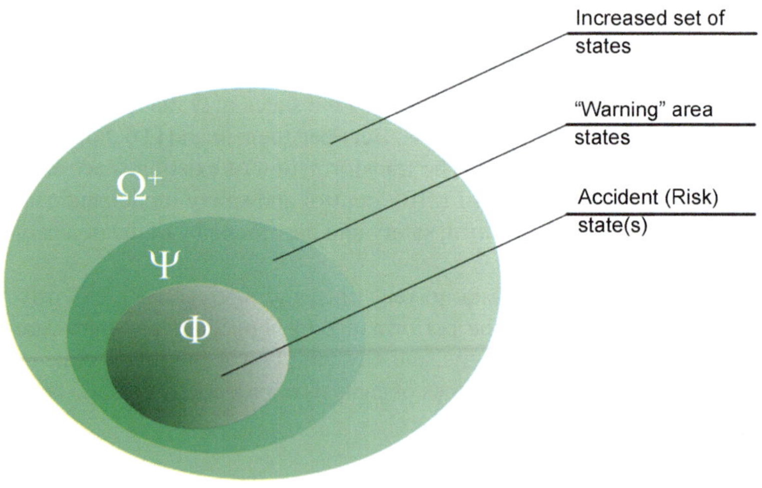

Fig. 11.6 The relation between different kinds of states

In other words, what we know (WWK) in relation with what does really exists: Ω. And the more we know the greater coverage. Knowledge can be gained in time; therefore, a coverage becomes a function of time; and not only time, but power of transformation, available information and gaining structural features of an object (again Structure, information and Time categories).

The Coverage, in fact, is a function of three independent variables: information, structure and time. Information here we mean as applied or aggregated in relation to the particular subject or of phenomena.

There is no doubt:

1. the coverage as a function of time and is either growing or at least not shrinking one
2. if the system and our knowledge about it evolves;
3. the coverage as a function of information is a growing function (the more we gather the more we can extract);
4. coverage as a function of structure is growing function—the more links and elements we can see and recognize, define in the system the more we know, at least statically.

All three functions coverage in time, coverage in information, coverage with model of structure is growing, assuming that—when system or knowledge evolves. All of the I, S, T growths cause growth on coverage and might have different gradients. The system we consider all growing, they all have different gradients.

It is worth mentioning that for any system description we can consider some invariant:

$$\text{System } = I \times S \times T$$

Overheads "to know everything about the subject" in terms of information, structure and time required is constant—we speak about Ω here (see above), not coverage.

For coverage, we can write:

$$\text{Coverage of System} = Ic \times Sc \times Tc$$

where all variables should be considered with "c" from the point of view of coverage.

11.7 System Software for Evolving Systems

System software for the next generation of evolving systems includes both the design of a new modular programming language called Active Language (AL), and two operating systems: a run-time system, called the Active Reconfigurable Run Time System (ARRUS) and a service operating system (SOS). AL and ARRUS are tightly coupled, as AL needs run-time system support for some of its features, such as recovery.

It is worth noting that ETHZ researchers such as N Wirth, J Gutknecht L Bläser and E Zouev were heavily involved in the domain of active language design over dozen years, while Vilenkin S and Schagaev I were designing and applying ARRUS prototype in 1988. This subchapter presents a kind of summary of our efforts in both segments.

11.7.1 Active Language (AL)

The language has to be able to describe active processes and is, in fact, a direct derivative of Oberon Wirth (1988, 1992) which includes the following new features:

1. A new data structure is introduced that eases the mapping of data to memory (dual string model);
2. GLM extensions will be exploited in the control and operator model;
3. Separation of interface and implementation to support dynamic software and hardware reconfiguration;
4. Deliberate reduction of the number of supported data structures and other language features;
5. Physical separation of constant, global and local variables and introduction of recovery points Kinnersley (2009);
6. Support for recovery and reconfiguration points at the module level using special program structures;
7. A special language feature that allows safe hardware access will be introduced which results in a completely type-safe language, without loopholes;
8. AL carefully revises dangerous language constructs, such as unbounded loops, and introduces the calculation of upper execution times and stack sizes to ease certification;
9. Rigorous memory management strives for an implementation without pointers and references.

An essential set of the above-mentioned concepts has already been implemented in our component-based, parallel programming language COMPOSITA Bläser (2006, 2007).

AL is a further development of this language, synthesizing additional concepts for real-time and fault-tolerance. In COMPOSITA, programs are entirely composed of active components that govern strict encapsulation and dynamic wiring, with a dual concept of offered and required interfaces and communication-based interactions (Fig. 11.7).

As the language is based on hierarchical composition and does not employ any ordinary pointers or references, surrounding components properly control the deletion of components and no garbage collection is needed for safe memory management Bläser (2024, 2007).

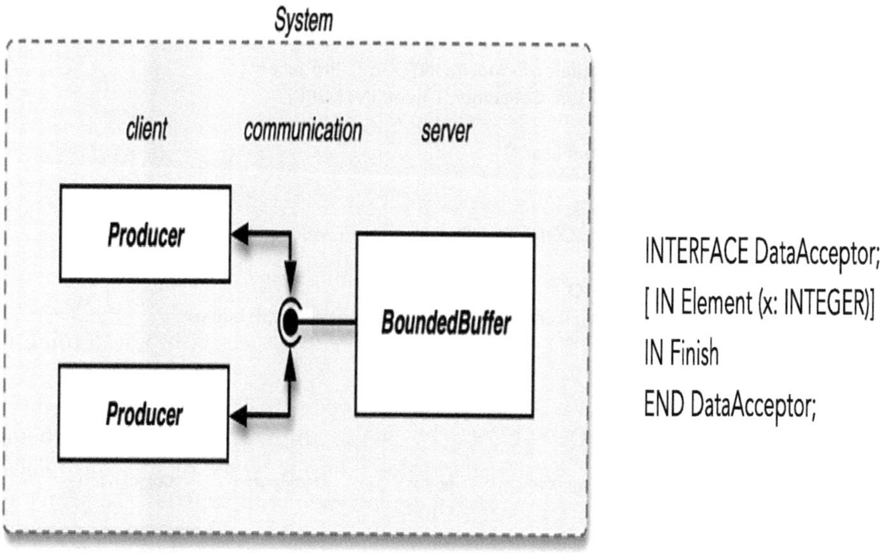

Fig. 11.7 Describing hierarchical component structures

In turn, Fig. 11.8 shows an example of a component structure, where a component can contain an inner network of sub-components. Communications follow a formal protocol written in an EBNF-like notation.

Due to the strict encapsulation, the components and can be easily mapped to various hardware architectures. Inherently, it enables parallelism (*N* components may be scheduled on up to *N* processors), as well as redundancy (the same components may be executed as multiple replicated instances). In subsequent work, a prototype compiler and run-time system for evolving systems has already been developed.

Direct support for reconfigurability and recoverability of program structures at the language level makes reconfiguration of system possible in case of hardware degradation due to faults, or task special requirements of system power-saving operation, or on the opposite, boosting task by using maximum hardware resources for completion of a task with required time limits.

For hardware fault tolerance, where malfunctions outweigh permanent faults, system software at the language level should include recovery points for the program at various levels of program presentation: procedure, module, and task. A detailed description of language support of hardware fault tolerance using recovery points is presented in Kaegi (2013).

Note here that the recovery point scheme will be embedded in the language and oriented towards the programs and data structure. This reduces the overhead for recovery after malfunctions and eases the impact of possible permanent faults.

```
COMPONENT System;
 VARIABLE
  buffer: BoundedBuffer; producer[i: INTEGER]: Producer;
  consumer[i: INTEGER]: Consumer; i, n, m: INTEGER;
 BEGIN
  (* read n and m from user *)
  NEW(buffer);
  FOR i := 1 TO n DO
   NEW(producer[i]); CONNECT(DataAcceptor(producer[i]), buffer)
  END;
  FOR i := 1 TO m DO
   NEW(consumer[i]); CONNECT(DataSource(consumer[i], buffer)
  END;
END System;
```

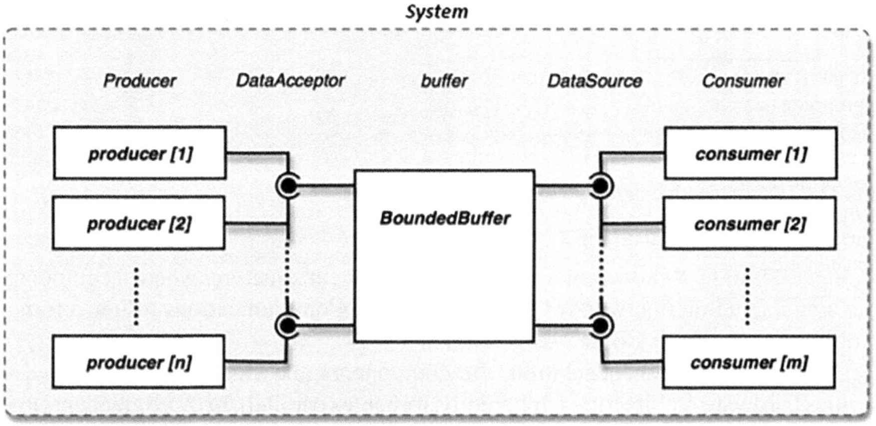

Fig. 11.8 Communication-based component

11.7.2 Active Reconfigurable Run-Time System

The run-time system, ARRUS, should also be involved in supporting real-time processing, as well as real-time reconfiguration of the underlying hardware elements and the respective network topology, including:

1. Flexible dynamic hardware resource management;
2. Software reconfiguration to adapt to changes in hardware states and system conditions;
3. Management of hardware/software interactions in the presence of hardware faults;
4. Hardware state monitoring and support of graceful degradation using testing and recovery procedures and reconfiguration management.

To match the required reconfigurability, parallelization, real-time, resilience to hardware degradation, and distributed control processing, the ARRUS itself is built in a strict hierarchical manner, as it is illustrated in Fig. 11.9.

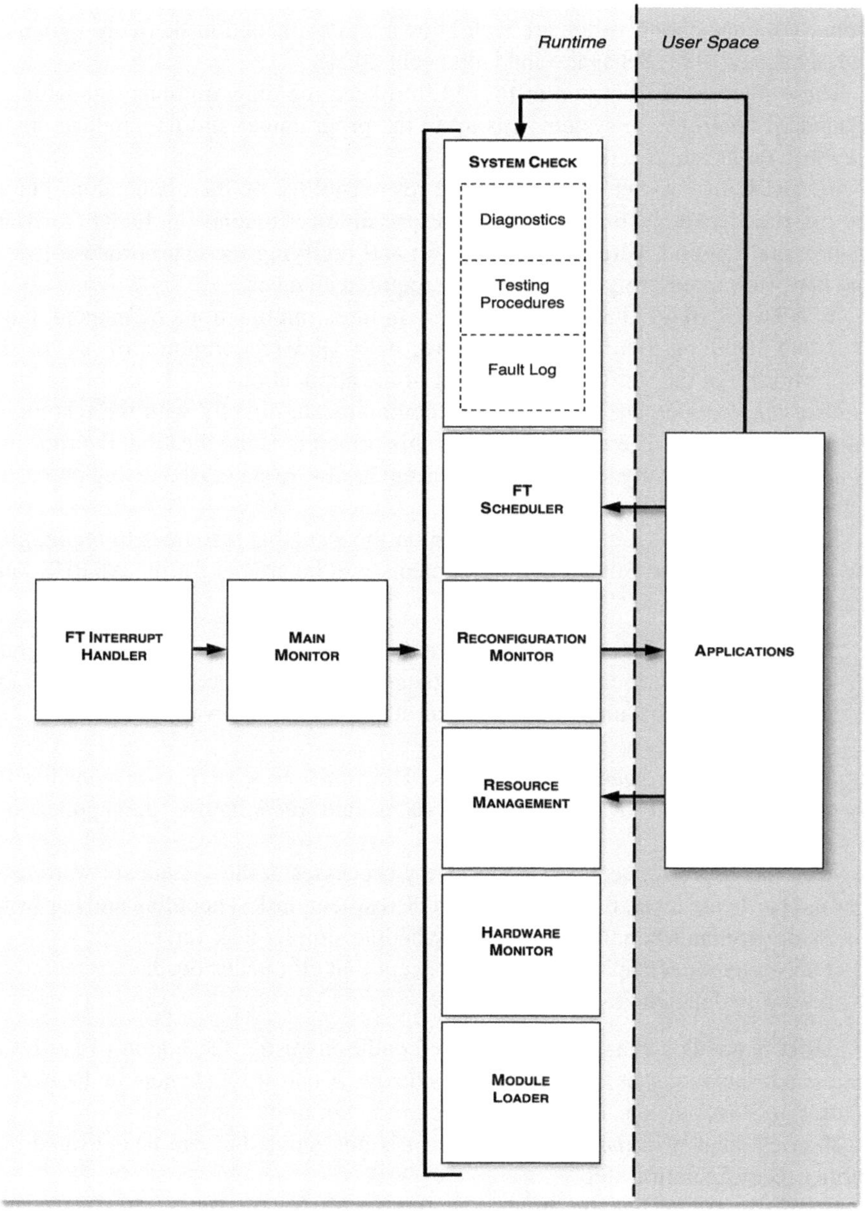

Fig. 11.9 Structure of run-time system for an evolving system

The lowest level module has no dependencies at all and consists of the main system monitor, which is responsible for the coordination of all activities, such as the initialization of reconfiguration entities, timer services (not shown), interrupt handling and all the remaining depicted functions.

ARRUS also provides all the standard functions of a run-time system, such as memory management, which are well known and explained in literature (Vilenkin and Schagaev 1988; Schagaev and Gutknecht 2024).

These features are omitted in Fig. 11.9 to keep the diagram understandable. In a standard control loop system, it is up to the programmer and the applications to diagnose faults and react appropriately.

In ARRUS, however, this is not the responsibility of the application, but of the run-time system. Thus, ARRUS is responsible for diagnosing faults (software failure, malfunction, permanent fault, etc.) and notifying the appropriate software and hardware monitoring services of any required changes.

In ARRUS, fault diagnosis (software failure, malfunction, permanent fault, etc.) and handling (fault elimination, recovery and reconfiguration) are in the responsibility of the run-time system, not of the application.

User applications are not allowed to communicate directly with the reconfiguration mechanisms. The rationale behind this principle is the idea that the run-time system is the only entity that knows the current hardware state, all ongoing processes and their resources.

In case of a fault, it can thus be based on the available resources to reconfigure the applications. The following developments will be crucial for the ARRUS fault handling mechanism:

1. Monitor hardware state. Possibly, either with interrupts (hardware signals changes), or periodic, software-initiated, hardware checking (further, both approaches are required to implement hierarchical hardware checking) Kaegi (2013);
2. Reaction on hardware state changes. The run-time system is responsible for the management of the hardware states and reconfigures the applications accordingly;
3. Collaboration of checking and recovery processes at the system software level and hardware level. Introduction of fault-resilient, task scheduling and hardware / software fault-handling strategies (schedule simpler task versions);
4. Fault-tolerant semaphores: A new concept that eliminates deadlocks caused by hardware deficiencies.

ARRUS handles changes to hardware conditions using the notion of hardware states and transitions, as Fig. 11.10 shows below. A hardware element in the Active, Master or Slave states is included in the current working configuration.

If an element is in the Stand-by state, it is not active, but can be activated in a further reconfiguration step.

If a fault is detected, the affected hardware element is set into the Suspected state which means that, at least for a while (until the generalized algorithm of fault

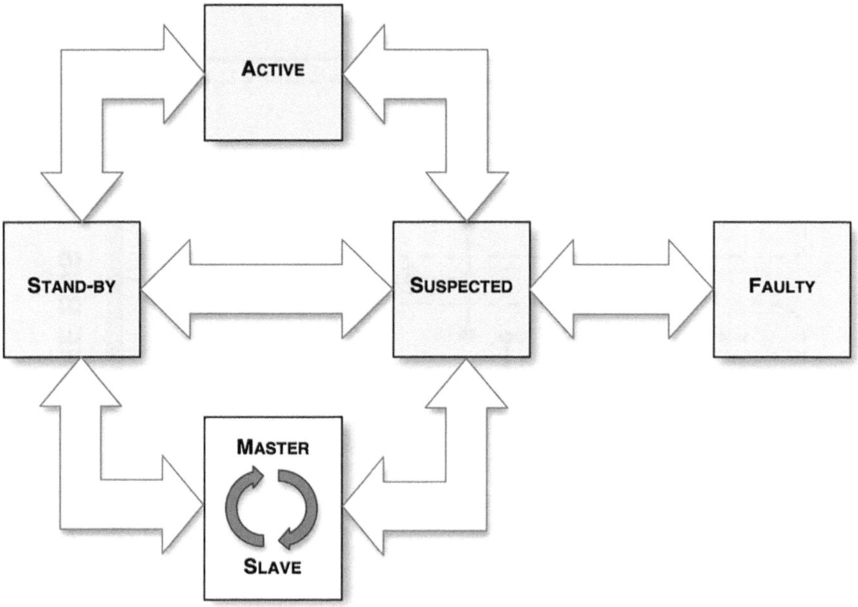

Fig. 11.10 Hardware states from the run-time system point of view

tolerance Schagaev (2008) is completed and a new valid configuration is established), this hardware element will not be included in any working configuration.

This representation of the hardware state for every ERA element defines the current configuration of the ERA hardware at the run-time system level. A configuration change can be triggered by changed application, power, reliability, or performance requirements, or a detected error.

11.8 Evolving System: Hardware

11.8.1 Basic Schemes

The indicative structure of a hardware element for an evolving system is presented in Fig. 11.11. Configurators called *T*-logic provide for flexible use of processor and memory elements in the system configuration related to performance and health conditions: when one processor is dealing with its own program (self-testing, autonomous calculations), it disconnects from the other nodes. The same technique is used to form reconfigurable hardware that is capable of adjusting to program requirements, or react to other events such as detected permanent faults.

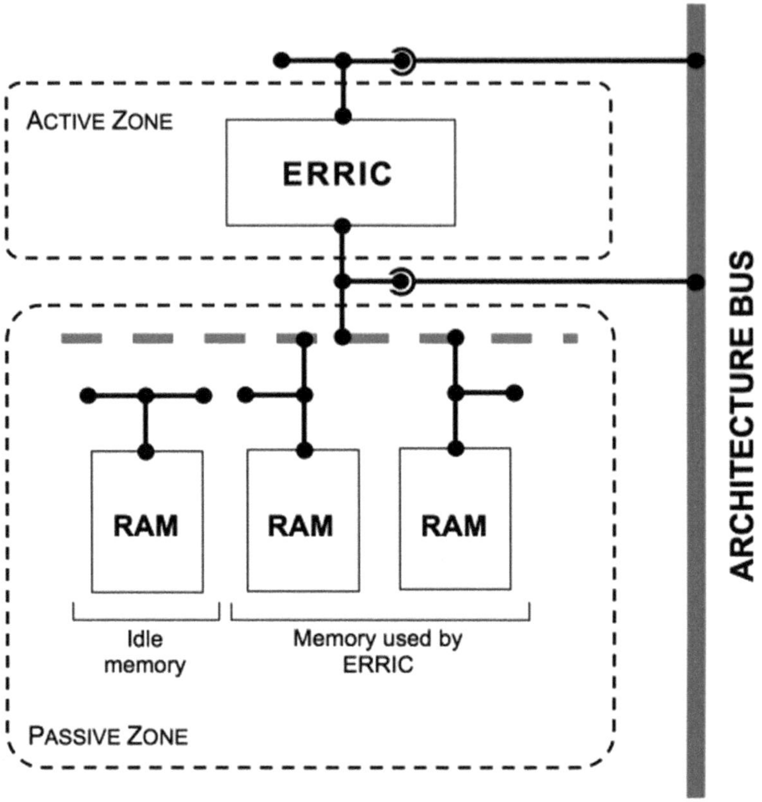

Fig. 11.11 ERA element

The memory configuration of an ERA element (Fig. 11.11) works for both: resilience, or performance. A special logic scheme, *T*-logic, configures the memory structure. When an application requires maximum reliability, the *T*-logic scheme might configure the memory as a 4-, or 3- (shown) unit with voter.

The configurations: two to compare and one spare, or three independent memory elements are possible. The number of memory elements might vary. For recent implementations, four memory elements were proposed. However, the principles of configurability by using *T*-Logic elements remain the same.

The instruction set of an elementary processor is designed to recover from hardware malfunctions by repetition of the instruction, making malfunction tolerance efficient Schagaev (2008).

In comparison with Motorola, ARM, and Intel, the proposed ERA is much simpler, with a higher level of parallelism and frequency able to be achieved. Due to simplicity-by-design, ERA only needs 10% power, when compared with the competitors, to reach the same clock speed.

The same technique is used to form a hardware configuration that is adjustable to program requirements, or when a hardware element itself (or architecture) detects hardware faults and cannot be involved in further program execution either on a temporary, or permanent basis.

The final decision about permanent isolation of an element is performed during a special mode of self–healing when testing and recovery procedures are executed.

Each element can be turned off individually to decrease power consumption. Note that the structure assumes only one leading element at a time, enforced by a "rotation" of the *T*-logic element. *T*-logic makes an ERA possible to operate until "the last man is standing," i.e. until a single processor, called ERA, and a single memory element can communicate.

By design, the Evolving, Recoverable, Reduced-Instruction Computer (ERA) is able to recover from major malfunctions and repeat the ongoing instruction when an error is detected. The bespoke instruction set and its declared top-down implementation of all required features reduces the impact of malfunctions on the performance and reliability of a system as a whole.

As it was shown in Schagaev (2008), the efficiency of malfunction tolerance depends on the fault coverage and the amount of involved redundancy. The efficiency of malfunction tolerance grows together with the ratio of malfunction to permanent faults.

An indicative performance comparison of Motorola, ARM and Intel to proposed architecture shows that the performance of ERA is not clock-to-clock competitive with Intel and ARM, but as the implementation is much (up to 6 times) simpler, a higher level of parallelism can be anticipated.

Additionally, early power estimation shows that ERA needs only 10% power compared to the competitors to reach the same clock speed. Finally, the absence by design of pipelining support and the ability to access memory at almost half that of, say, ARM, enables system monitoring and processor frequency variance when applications require extreme performance.

Below, Fig. 11.12 shows a first prototype of ERA, designed and developed by the authors in collaboration with ITACS Ltd.

Memory, organized as four, mutually replaceable schemes with virtual addressing configures the ERA for Reliability (two pairs, or triple structure with spare element), Energy-smartness (one element is active, the rest disabled), or full capacity—when all elements form one memory bank for the program. The ROMs function as one, or both can be active. At the moment, all reconfigurations and ERA processor elements (four of them) are assembled on Altera PLD, shown.

Further development will obviously lead to special chip for processors, memories with reconfiguration support and specula design of T-logic element to achieve it maximum simplicity, performance and reliability.

The proposed reconfiguration schemes and design approach is the subject of further development, as it provides a unique linear exchange, or reliability/ performance and power consumption, dependent on the actual need and request of user task, or run-time system.

Fig. 11.12 ERA element—first prototype

11.9 Evolving System: Multi-element Configuration

There is no doubt—see "model of rope and box" above - one has to address how, if possible, we can use the resources of multi-agent systems supported by multi-element architecture. Conceptually, Fig. 11.13 presents the next step for ERA as a prototype of evolving systems.

Each element of the proposed architecture might serve as a node in an evolving system: it can be turned off individually to decrease power consumption, or made inactive yet able to be included in any application configuration, or aggregated with others for maximum capacity or reliability.

To recap, when applied throughout the whole system, T-logic elements make the whole evolving system operate "until the last man is standing," i.e. until a single processor, ERA, and a single memory element can communicate through the remaining links.

Note that active and passive elements (processor and memory) can be at the extremities in this scenario.

Various areas of Evolving Architecture can be involved in different types of tasks. Some segments might run safety–critical tasks—extremely demanding in terms of reliability, while others execute heavy calculations of matrix algebra.

The pink-coloured part of the scheme illustrates that when, for any reason, the Active zone (ERA) is not in action (either because of task configuration, or because a permanent fault has induced its idle status by system software).

Other processors might use local memories of one hardware section within ERA.

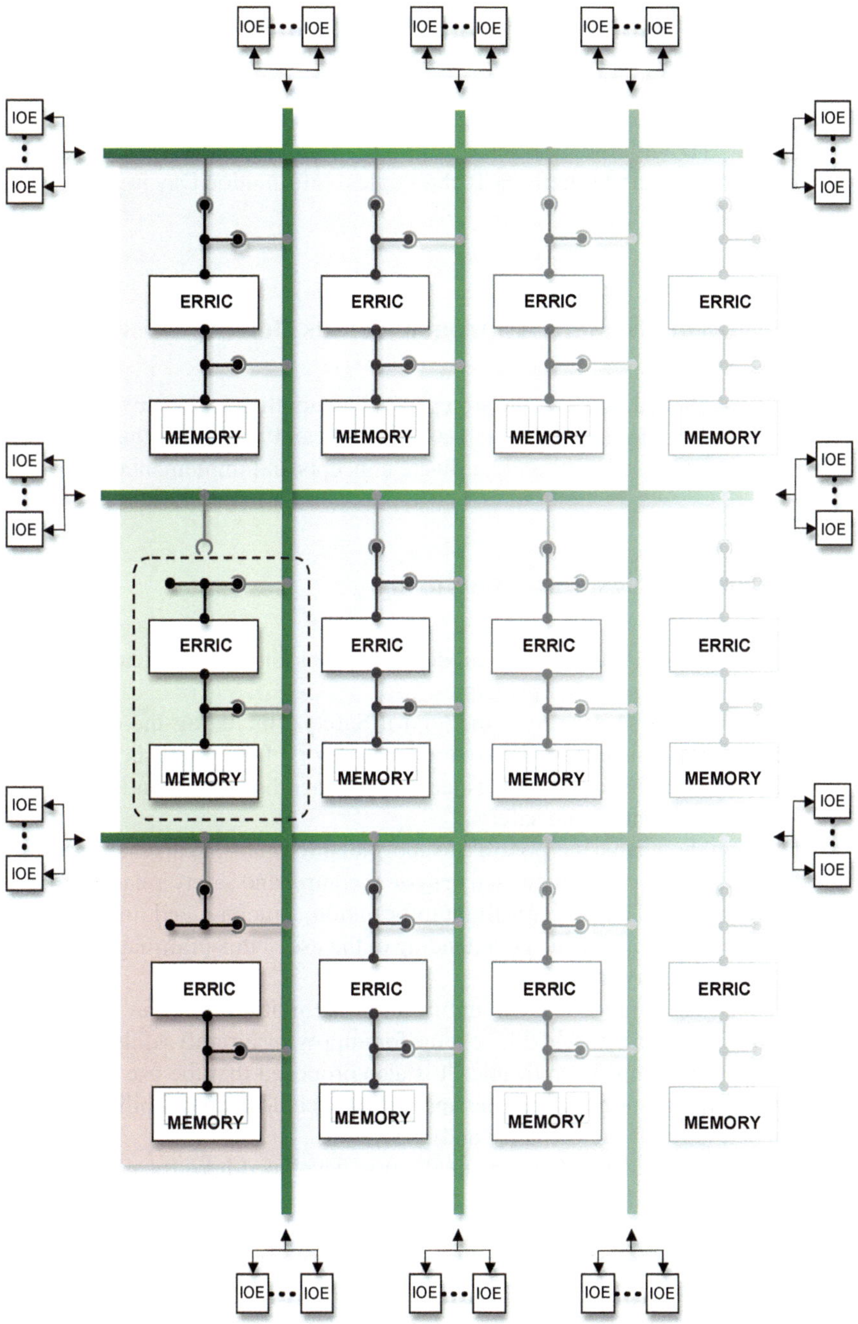

Fig. 11.13 Indicative architecture of evolving system

Finally, the dotted part of the scheme illustrates that the leading active element has only one connection, as it is counted as active and initiating an exchange; the others involved are also "listening".

The proposed approach is promising. Initially it was presented by Professor Schagaev IT-ACS LTD with the assistance of Professor Gutknecht (ETH Zurich) at EC program FET (Future and Emerging Technologies) Information Day in 2005 January 14th, Brussels.

11.10 Evolving System Approach Versus Berkley View

The recently published, Berkley view on parallel computing Asanovic (2006) differs from our approach. Table 11.4 is self-explanatory and illustrates the differences, showing both system software and hardware concepts and implementations.

11.11 Evolving System: Conclusion

1. In this Chapter the theoretical model of active system control and safety has been developed from first principles.
2. The Principle of Active System Control Safety—including theoretical background and its connection with redundancy theory, fault tolerance and dependability has been introduced. This leads to a generalised information model for active system control and safety.
3. It has been shown that it is possible to apply all three main agents (hardware, software, and user) to implement active system control and safety-related processes and algorithms. This requires flight information, structure, and time dependencies to be included and an understanding of the use of this information for active control, and safety monitoring.
4. The use of Information and Communication Technology provides the best basis for introducing an unavoidable scheme for improving aircraft safety and making active system control achievable. It is also proposed that the use of hardware/software safety monitors can be applied to exclude the possibility of cutting corners in aircraft maintenance and safety.
5. The application of PASC in the medium is developed here for the analysis of states of aircraft during and after flight.
6. The feasibility of the active system safety approach has been demonstrated, and thus prevention of some accidents in real time of flight is possible. Various mathematical models and their features are briefly analysed.
7. The nature of the connections between flight data, elements of an aircraft, the fault dependencies between them and performance of flight data analysis should be considered and will be explored in further chapters.

Table 11.4 Berkley view versus evolving reconfigurability approach

Berkley view	Evolving reconfigurability approach
• Seven goals (and 11 bullet points) • Future computers must be effectively parallel • It is possible to consider 1000 + cores • Performance measurement for parallel computing should be reevaluated and new metrics introduced	• New rigorously defined computing paradigm based on GLM exploits automatically maximum possible parallelism of both: algorithm and hardware available and minimize synchronization complexity at the run time to reduce concurrency. This approach guarantees together with flexibility of hardware configuration for both: performance or reliability purposes. Adjust available hardware for maximum possible efficiency across the whole broad class of applications usually covered by semantically different architectures such as VLIW, SIMD, MIMD
1. Auto-timing of software and hardware 2. Human centric computing in multi-core 3. Application of wide range of data typing	• Software will be written in traditional way, and debugged on a standard systems • Parallelization of the algorithm will be done by backward compilation of a sequential program. Using GLM again for representation of task potential parallelism and fine-tuning of hardware resources available on the wafer, before and during application runs
1. Three levels of parallelism should be pursued: task level word level and bit level 2. Parallel programs must be presented independently to the number of processors available 3. Limitations on features that reduce parallelism 4. OS functionality should be based on libraries and virtual machines	• Task level parallelism is prerogative of run time system and heavily dependent on workload and available resources during application run: therefore, dynamic scheduling to optimize task parallelism is exception not the rule in evolving architecture. For evolving reconfigurable architecture dynamic support of parallelism should be an exception not a rule • Application of GLM for both redesign of algorithms and timing on architecture available resources (supported by *T*-configurator at the element and architecture levels) maximize parallelism and minimize concurrency
1. A higher-level rate of fault is expected for multicore systems	• Assumptions about higher rate of permanent hardware faults for the next generation of electronics are not correct. More details see, for example, in Feynman lectures…
1. SEC/DED options proposed	• Number of malfunctions caused externally by alpha particles and internally due to higher density of elements on the wafer does not lead to increased reliability by using SEC/DED; most likely 16 + bit errors will take place in hardware. System software contribution to the malfunction of the system caused by support of dynamic parallelism and complex concurrency monitoring
1. Synchronization overheads should be reduced	• Overloaded by task parallelism, the monitoring OS will be deadlocked. We propose a reconfigurability of available hardware and tasks by re-compilation of existing algorithms

(continued)

Table 11.4 (continued)

Berkley view	Evolving reconfigurability approach
Wide range of data types should be: 1. 1 bit (Boolean) 2. 8 bits (Integer, ASCII) 3. 16 bits (Integer, DSP fixed point, Unicode) 4. 32 bits (Integer, Single-precision FP, Unicode) 5. 64 bits (Integer, Double-precision FP) 6. 128 bits (Integer. Quad-Precision FP 7. Large integer (>128 bits))	• We propose Dual string data structure where for each data element of the array a special descriptor defines data type (altogether 232 types) • This covers all possible data types user can dream or imagine Efficiency of access to the proposed data structure is equal to the array access
1. New models should support proven styles of parallelism 2. FPGA systems are future HW platforms for multi-core computing	• Styles of parallelism are application specific and vary due to technology modification. We propose an auto-tuning of existing programs into their maximum parallel form • FPGA technology at the element level and specially designed wafer with pre-fabricated configuration fabric of active processing element and passive elements enables monitoring architecture for performance, power consumption and reliability

8. This work presents concepts and design ideas on how to cope with some known drawbacks of computer architectures. Evolving systems of future computers will exploit reconfigurability of hardware and software for Performance, Reliability, or Energy-wise functioning.
9. At the system software level, a reconfigurability of hardware and software should be introduced, represented and supported by programming language, service operating system and run-time system. Ultimately, this will result in a simple, yet scalable, reliable system and provide performance and power saving options, with linear trade between.
10. To avoid existing drawbacks of parallel and other reconfigurable architectures, several holistic principles were presented and pursued through the whole life cycle of computer systems: from the preparation of algorithms down to the execution of programs and hardware.
11. We presented a new approach, aiming to introduce a system with evolving features, based on the mutual design of architecture and system software. System software support for hardware reconfigurability is centered at the compiler level, leaving only essential reconfigurability handling at the run-time level.
12. It was shown that consistent support of system reconfigurability eases parallelization, reduces concurrency, assists fault tolerance and implements a power-awareness for applications, when necessary. Two models: Control, Data, Predicate dependencies of the program and the Graph Logic Model represent concurrency and parallelism of a program and enable their explicit separation. A

sequence was proposed to find the minimum form of any program in terms of time overheads, limited only by available hardware resources. The generalized algorithm of parallelization of sequential programs was presented.
13. The structure of a reconfigurable, run-time system was discussed, together with a scheme for hardware configurability support.
14. Hardware elements and the structure of the whole architecture were presented, with explanations of how maximum reconfigurability ("'til the last man is standing") is achieved.
15. A comparison of the evolving reconfigurability approach with known approaches was presented. In contrast with Von Neumann, we are attempting the design of reliable (and reconfigurable) systems with reliability of components.

References

Ames B (2007) Intel tests chip design with 80-core processor. http://www.macworld.com/news/2007/02/12/intel/

Asanovic K, Bodik R, Catanzaro B, Gebis J, Husbands P, Keutzer K, Patterson D, Plishker W, Shalf J, Williams S, Yelick K (2006) EECS Department, University of California, Berkeley Technical report No. UCB/EECS-2006-183 http://www2.eecs.berkeley.edu/Pubs/TechRpts/2006/EECS-2006-183.pdf

Berka S (2009) Operating system documentation project. http://www.operating-system.org/betriebssystem_english/index.htm

Blaeser L (2024) Modern runtime system and compiler design book (Luc Bläser) Independently published ISBN 978-3-033-08044-7. Copyright (c) by Prof. Dr. Luc Bläser

Blaeser L. A programming language with natural persistence Luc Bläser 2006/10/22 Buch companion to the 21st ACM SIGPLAN symposium on object-oriented programming systems, languages, and applications, pp 637–638

Bryant R, O'Hallaron D (2002) Computer systems: a programmer's perspective. Prentice Hall

Dijkstra EW (1965) Solution of a problem in concurrent programming control. Commun ACM 8:569

Everett R, Swain F, Report R-127 (1947) Whirlwind I computer block diagrams. MIT Servomechanisms Laboratory

Fisher JA (1983) Very long instruction word architectures and the ELI-512. In: Proceedings of the 10th annual international symposium on computer architecture. ACM, Stockholm, Sweden, pp 140–150

Flynn M (1972) Some computer organizations and their effectiveness. IEEE Trans Comput C 21:948

Goldstine HH, Goldstine A (1982) (1946) The electronic numerical integrator and computer (ENIAC). The origins of digital computers: selected papers. Springer, New York, pp 359–373. ISBN 978-3-540-11319-5

Gosden J (1966) Explicit parallel processing description and control in programs for multy-and uni-processor computers, AFIPS'66 (Fall): proceedings of the November 7–10, ACM, New York, pp 651–660

Gutknecht J (2006) The dining philosophers problem revisited. JMLC, Lecture Notes Comput Sci 4228:377–382. Springer, Berlin

Hennesy J, Patterson DA (2025) Computer architecture a quantitative approach. ISBN: 9780443154065 978-0-443-15406-5 eBook ISBN: 9780443154072

Hofstra University (1999) History in the computing engineering curriculum. Appendix 4:1950–59. www.comphist.org/pdfs/CompHist_9812tla4.pdf

Holland JH (1960) Iterative circuit computers, western joint IRE- AIEE-ACM computer conference, May 3–5, 1960, San Francisco, California, pp 259–265, ACM, New York, USA
IBM (2007) Power 6 specs: IBM Power6 microprocessor and IBM system, p 570
Kaegi-Trachsel T, Schagaev I, Gutknecht J (n.d) (2013) Hardware testing on the level of tasks. In: 30th IFAC workshop on real-time programming and 4th international workshop on real-time software. Presented at the international multiconference on computer science and information technology
Kinnersley B (2009) The language list. http://people.ku.edu/~nkinners/LangList/Extras/langlist.htm
Lamport L, Melliar-Smith P (1985) Synchronising clocks in the presence of faults. J ACM 32(1):52–78. ACM, NY
Lamport L (1983) The weak Byzantine generals problem. J ACM 30(1):668–676. ACM, New York
LaPedus M, EETimes (2003) Intel gears up 90-nm processor, chip set rollout. http://www.eeEmes.com/conf/idf/showArEcle.jhtml?arEcleID=10800811&kc=3172
Monkman S, Schagaev I (2013) Redundancy + repeatabilit = recoverability. Electronics 2:212–233. https://doi.org/10.3390/electronics2030212. ISSN 2079-9292
Nair R (2002) Effect of increasing chip density on the evolution of computer architectures. IBM J Res Dev 46:223–234
Newell A (1960) A on programming a highly parallel machine to be an intelligent technician, IRE-AIEE-ACM'60 (Western), May 3–5, ACM, New York, pp 267–282
Plyaskota S, Schagaev I (1995) Economic effectiveness of fault tolerance. Autom Remote Control 07:1017–1026
Schagaev I (2008) Reliability of malfunction tolerance. Proceedings of the international multiconference on computer science and information technology, pp 733–737. ISBN 978-83-60810-14-9. ISSN 1896-7094
Schagaev I Gutknecht J (2024) Software design for resilient computer systems. https://doi.org/10.1007/978-3-031-55139-0 ISBN 978-3-031-55138-3, Published: 16 July 2024
Schagaev I, Kirk B (2025) Active system control, Springer ISBN 978-3-031-77180-4. Published: 22 January 2025
Schagaev I, Zouev E, Kaegi T (2020) Software design for resilient computer systems. Springer AG. https://doi.org/10.1007/978-3-030-21244-5
Schwartz E (1961) An automatic sequencing procedure with application to parallel programming. JACM 8(4):513–537. ACM New York, NY, USA
Slegel TJ, Averill RMI, Check MA, Giamei BC, Krumm BW, Krygowski CA, Li WH, Liptay JS, MacDougall JD, McPherson TJ, Navarro JA, Schwarz EM, Shum K, Webb CF (1999) IBM's S/390 G5 microprocessor design. IEEE Micro 19:12–23
Slotnick DL et al (1962) The SOLOMON computer, AFIPS'62 (Fall) proceeding December 4–6, 1962, Philadelphia, ACM, New York, pp 97–107
Squire JS, Palais SM (1963) Programming and design considerations of a highly parallel computer. AFIPS Spring joint computing conference, pp 395–400
Turski WM, Wasserman AI (1978) Computer programming methodology. SIGSOFT Softw Eng Notes 3:20–21
Vilenkin S Schagaev I (1988) Operating system for fault tolerant SIMD computers programmirovanie, vol 3
Whirlwind I (1952) Master drawing list and general rack layout of computer. MIT, Department of Electrical Engineering, Cambridge
Williams HC (2003) The writing on the wall. University of Pensilbvaniya Press. ISBN 0-8122-3711-0
Wirth N (1988) The programming language Oberon. Softw Pract Exper 18:671–690
Wirth N (1983) Programming in modula-2. Springer, TELOS
Wirth N (1992) Project Oberon: the design of an operating system and compiler. Addison-Wesley Publication (Sd)

Chapter 12
Some Further Notes: Degree of Freedom

Abstract Quite a few of my respected colleaguesяы from Canada, UK, USA, France, Germany, Switzerland were asking how my redundancy theory and reconfigurability concept might be applied, if ever. I have found nothing else but to present some examples and demonstrate what reconfigurability and recoverability can do in concert. Primarily by providing flexibility for applications by a specially designed structure of a system. Usually in engineering it was called a bit more generic term: degree of freedom. For our systems, the degree of freedom (DoF) is not limited by freedom of motion in space; some examples elaborate this.

Keywords Degree of freedom · Performance · Reliability · Energy efficiency · Reconfigurability · Evolving architecture

12.1 Bits on Degree of Freedom

Speaking about DoF, we should not limit ourselves by motion, concern only three, two or one dimensions.

In fact, every function defined in the system as main (uninterrupted shooting, precision, targeting, reliability, etc., for machine gun—to name one) can be described in terms of degree of freedom or levels of operability.

Clear, that level should be preserved as high and as long as possible, here it is worth to read in our previous books—as an example of trade performance versus reliability: (Schagaev et al. 2020; Castano and Schagaev 2015).

Let's check it out in a bit more detail.

We are aware that the more energy we can implement in the system, the more performance we might expect. But we understand that there is a natural limit in performance gain, defined by technology, system design, and even assembling quality.

We are also a kind of aware that performance and reliability are not great friends: 1600cm3 engine of Formula 1 provides 1600hp, but the mean time to failure (MTTF) for this type of engine is limited by a few races.

© The Author(s), under exclusive license to Springer Nature Switzerland AG 2026
I. Schagaev, *Resilience of Computer Systems*,
https://doi.org/10.1007/978-3-032-00938-8_12

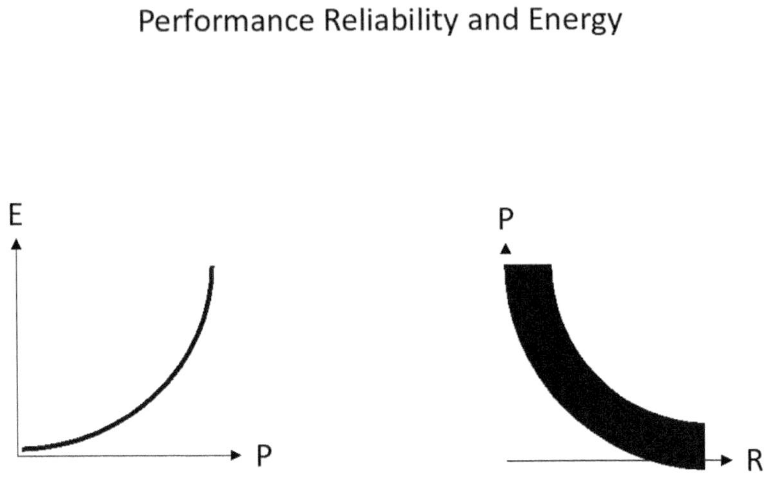

Fig. 12.1 Energy versus performance, reliability versus performance

At the same time 40hp engine of VW1200 1965 made—from my own experience, provided MTTF 30 (thirty!) years, while gearbox GUARANTEES functioning over half a million kilometres.

To remember these dependencies of performance and reliability, a qualitative Fig. 12.1 might be helpful.

The forms of curves might vary a bit but it is clear that more energy might not bring more performance, while bringing more problems with reliability.

Speaking about system we design—we naturally expect that within given limits of performance and reliability it operates or it should operate fine.

Reliability, it is clear that between 100% and acceptable level for every system should be obligatory considered together with performance of a system.

We can say that at the first moment of operating of the system, by design, it provides 100% performance at 100% reliability, but next moment, we have to accept that both might deteriorate and at different rate.

Physics, physiology (when human is involved in the system as a part of it)—all interfere. Therefore, we have to think about acceptable margins of reliability and performance and about an "envelope of trade", as Fig. 12.2 illustrates.

Thus, we are facing an envelope of performancereliability trade (grey segment), where all acceptable states of a system reside. It is also clear that systems we design do not have differentiable number of states—workable states are counted and numbered.

All of them reflect a level of degradation, or predefined configuration, that is changeable due to changing of conditions, and supported by configurations we prepare, if any.

Fig. 12.2 Performance—reliability margins and envelope

It is obvious that best, fairest trade of "performance versus reliability" should be linear—say 10% of performance decrease should bring 10% gain of reliability—from top left corner of rectangular down to bottom right corner, all states should be placed. But it is highly unlikely achievable.

One might ask: "OK, sounds good, did you try yourself to implement your theory?".

Actually, we did. Our on-board active system control computers (Fig. 12.3) made in 1991–94 have passed field testing, showing ability to degrade in reliability while preserving functioning, reducing performance for a sake of reliability at unprecedented level.

The main computer was capable to work at a quarter of board—disabling all not required for current mission elements and using a fraction of energy—a special mode, working at full speed on one board and doubling either performance or reliability while using both boards.

In turn, system memory—that control functioning of the rest of system and was accessible only by operating system, (Fig. 12.4), had 3–2–1 stages of reliability degradation as well as 1–2–3 memory capacity—when required.

More details about this design see (Schagaev and Buhanova 2001; Bernstein et al. 1992), further development at the die level is leave to future engineers. Mission memory, Fig. 12.5, had exceptional reliability, supported by predefined and designed reconfiguration, while 32,7,2 bits configuration was capable go down to 32,1 and 32,0 bit configuration preserving operational data, sacrificing reliability, increasing recovery time and requiring a system software support.

Overall reliability of our system exceeded by the order of magnitude all known aerospace designs and since then still unachievable by known systems. More 3 4

Fig. 12.3 PRE-smart active system control computer

Fig. 12.4 Reconfigurable for reliability and performance triplicated memory

Fig. 12.5 Mission memory and processor of PRE-smart system

details about our mission memory design can be see and further elaborated in our book published in (Schagaev and Buhanova 2001; Bernstein et al. 1992).

Figure 12.5 shows a structure of one board from our system; we are blurring details of elements for obvious reasons.

Thus, theory and practice of introducing in the system design solutions for preserving degrees of freedom and enabling trade of performance, reliability and energy is feasible and suggests an effective way of design.

Figure 12.6 conceptually presents how from system design of fault-tolerant system we came to the idea of system with the ability of P-R-E trading.

This approach is not limited by ICT systems and might be procrastinated in the domains of human activities and projects.

Example how we came to the idea of PRE-smart trading was fully explained by us in all referenced here publications (Schagaev et al. 2020; Castano and Schagaev 2015; Schagaev and Buhanova 2001; Bernstein et al. 1992; Bernstein and Tomfeld 1993).

Another design, illustrated by Fig. 12.7 shows that trading at system level can be pretty flexible, memory banks can be reconfigured for the purposes of reliability and maximum capacity using special supportive fault-tolerant logic.

Figures we assume are self-explanatory. Further details are available from our books mentioned above.

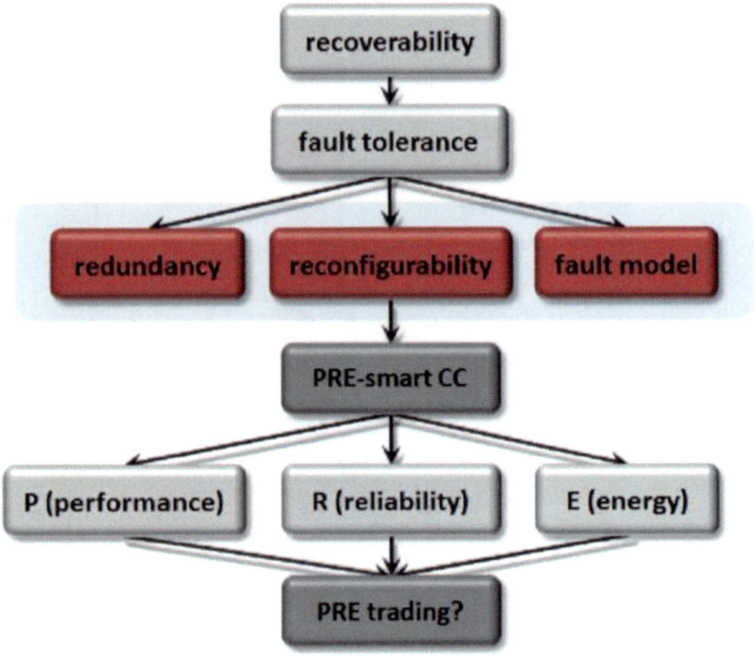

Fig. 12.6 Evolution of trading for fault tolerance to PRE-smart system design

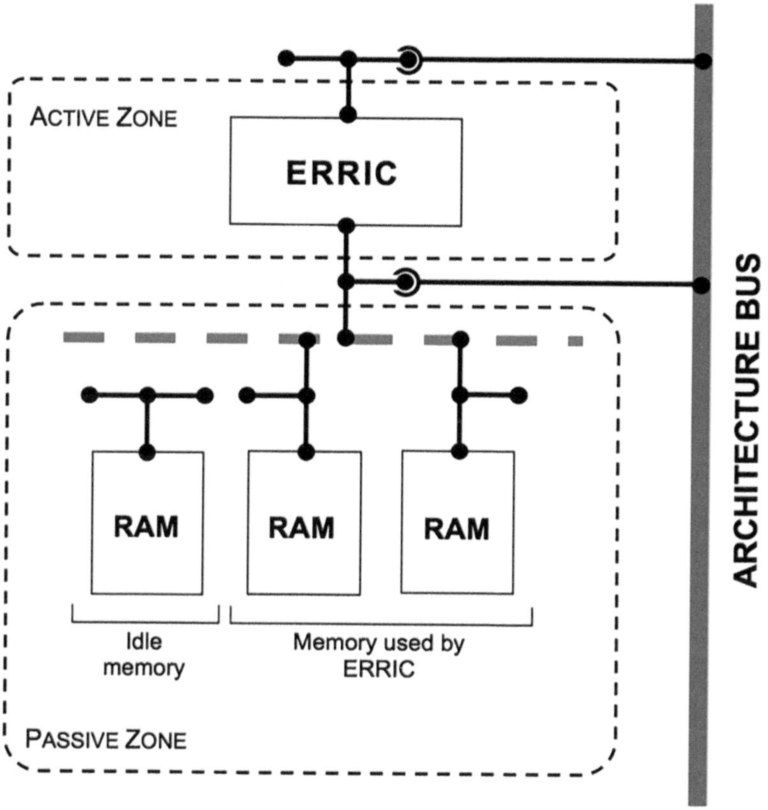

Fig. 12.7 Trading energy, capacity and reliability

References

Bernstein A, Tomfeld Y, Schagaev II (1993) High reliability RAM-II, automatic and remote control. Pergamon press, pp 1012–1021

Bernstein A, Tomfeld Y, Schagaev I (1992) High reliability RAM-I, automatic and remote control. Pergamon press, pp 449–455

Castano V, Schagaev I (2015) Resilient computer system design. Springer. https://doi.org/10.1007/978-3-319-15069-7

Schagaev I Buhanova G (2001) Comparative study of fault tolerant RAM structures proceeding IEEE dependable systems and networks conference Goteborg, Sweden

Schagaev I, Zouev E, Kaegi T (2020) Software design for resilient computer systems. Springer AG. https://doi.org/10.1007/978-3-030-21244-5

Index

A
Assembler, 197
Availability, 1, 9, 12, 23, 29, 31, 33–37, 39–41, 88, 92, 161, 209, 229, 231

C
Checking, 51, 65–68, 78, 82, 89, 115–117, 127, 128, 130, 131, 134, 150, 160, 166, 170, 172–178, 203, 211, 219, 226, 238
Concurrency, 186
Coverage, 5, 32, 33, 39, 60, 75, 76, 78, 79, 82, 92, 106, 109, 110, 116, 119, 130, 134, 171, 230, 232, 233, 241

D
Debugging, 197
Degree of freedom, vi, ix, 249
Dependability, 11
Dynamic Random Access Memory (DRAM), 47, 80, 94, 184

E
Embedded systems, viii, 1, 2, 4, 67, 96, 112
Energy efficiency, 101, 119
Energy trade, 1
Energy-wise systems, 215
Error detection, 50, 57, 58, 62, 65, 67, 73–75
Evolving architecture, 242
Evolving reconfigurable architecture, 5, 211, 229
Evolving systems, 220, 233, 235, 242

F
Fault tolerance, 2, 5, 48, 91, 101, 114, 210

G
Generalised Algorithm of Fault Tolerance (GAFT), 119

H
Hardware design, viii, 5, 209, 224
Hardware prototype, 203

I
ICT, viii, 124, 217, 218, 253
Informational redundancy, 208
Instruction execution, 198–200
Instruction set, 135
Interaction, 181
Interface logic, 123
Interruption, 181, 190

M
Maintainability, 29, 30, 40, 105
Malfunctions, 2, 134, 166, 176, 188, 219, 235, 240, 241, 245
Memory, 70, 129, 132, 135, 139, 153, 156, 163, 165, 205, 211, 241
Model of fault, 92
Monitoring of hardware configuration, 169
Mutual exclusion, 183, 188, 192, 211, 223

O
Operating systems, 218, 220, 233

© The Editor(s) (if applicable) and The Author(s), under exclusive license to Springer Nature Switzerland AG 2026
I. Schagaev, *Resilience of Computer Systems*,
https://doi.org/10.1007/978-3-032-00938-8

P

Performability, 37, 38
Performance, vii–ix, 1, 3–5, 9, 13–15, 33, 38, 39, 41, 49, 57, 67, 71, 72, 82, 87, 92, 101, 111, 112, 116, 119, 123–125, 129, 136, 141, 142, 146, 148, 152, 156, 166, 173, 180, 184, 185, 187, 188, 201, 209–212, 215–220, 223, 224, 229–231, 239–241, 244–246, 249–253
Permanent faults, 73, 74, 89, 117, 159, 170, 171, 219, 235, 239, 241
Probability, 12, 15–18, 21, 23, 28–32, 34, 35, 37, 38, 48, 69, 78, 94, 119, 161, 207, 211, 231, 232
Processor, 116, 123, 131, 138, 151

R

Radiation, v, viii, ix, 1–5, 47, 55, 89, 90, 95, 96, 128, 188, 210
Real-time, 1, 72, 82, 112, 166, 174, 179, 187, 188, 210, 231, 234, 236, 237
Reconfigurability, 125, 146, 166, 219
Recoverability, 28–31, 36, 37, 39–41, 97, 102, 105, 110, 111, 117, 132, 166, 210, 235, 249
Redundancy, 25, 48–50, 54, 57, 81, 111, 125, 219
Reliability, vi–ix, 1–5, 9, 10, 12–28, 31, 33, 34, 36–42, 45–48, 53–57, 61, 62, 68, 87, 88, 91, 92, 95, 101, 110–113, 116, 119, 123–125, 128, 130, 135, 136, 143, 146–150, 152, 154, 156, 160, 161, 165, 166, 179, 180, 198, 203, 204, 206–211, 215, 217, 219, 221, 222, 225, 226, 230, 231, 239–242, 245–247, 249–253, 255
Reliability model, 87
Resilient architecture, ix, x, 123

S

Safety, vii, 28, 29, 40, 41, 244
Safety-critical, vii
Simulator of architecture, 197
Static Random Access Memory (SRAM), 71, 80, 204, 205
Structural redundancy, 50, 75, 78, 81, 82, 111, 112, 120
System failure, 9, 10
System failure life cycle, 9

T

Testing, ix, 5, 20, 22, 47, 55, 60, 62, 81, 89, 105–107, 110, 154, 164–166, 169–173, 175, 176, 179, 206, 211, 212, 236, 239, 241, 251
Time redundancy, 50, 72–78, 80, 82, 92, 208

GPSR Compliance

The European Union's (EU) General Product Safety Regulation (GPSR) is a set of rules that requires consumer products to be safe and our obligations to ensure this.

If you have any concerns about our products, you can contact us on ProductSafety@springernature.com

In case Publisher is established outside the EU, the EU authorized representative is:

Springer Nature Customer Service Center GmbH
Europaplatz 3
69115 Heidelberg, Germany

Batch number: 10105676

Printed by Printforce, the Netherlands